WINNING
THE HOPES

REFLECTIONS FOR A BRIGHTER FUTURE

TRONG NGUYEN

Trong Nguyen

WINNING THE HOPES:
REFLECTIONS FOR A BRIGHTER FUTURE

Copyright © 2020 by Trong Nguyen

The stories in this book are all based on true events. However, certain scenes, characters, names, businesses, incidents and locations have been fictionalized to enhance the storytelling and illustrate key academic concepts and practical lessons learned.

For information contact: trong@winningthehopes.org

Editing by Cabrina Attal and Kathy Clolinger

Cover Design and Illustrations by Pia Reyes

ISBN: 978-0-9985702-5-9

First Edition: February 2021

DEDICATION

∽

This book is dedicated to two kind souls from St. John's, Newfoundland, Canada – **Loretta** and **Shawn Dobbin**. Their acts of generosity forever changed the lives of a family in need. Thank you for being the inspiration for us to pay it forward!

Trong Nguyen

TABLE OF CONTENTS

FORWARD

The pandemic of 2020 was a wake-up call to everyone around the world. For the first time that our generation can recall, we were told to stay away from friends, family, and loved ones because that was the only way to stay safe and alive (literally). Lockdowns and "shelter in place" became our new normal. Isolated, I had a lot of time to think and reflect. This book is a bit different from my first two. It's not a how-to-sell book and it is not a fun, deal-chasing book like *Winning The Bank*. Instead it is an introspective look at and reflections on my past experiences with commentary on what I think is important.

Maturity, Catholic guilt, mid-life crisis, the coronavirus. Maybe it was a combination of all four that started me down this path of wanting to give back and make a difference. The enlightened ones realize early on that the most meaningful thing you can do while you are here is to help others. It took me the longest time to realize and appreciate that fact. Better late than never, I guess.

Winning The Hopes Inc., is a non-profit that I started with a mission to provide aid and mentorship to children in need. Our goal is to help these kids so that they can get the education, training, and coaching they need to obtain a post-secondary education. Hopefully this will change the trajectory of their lives and provide them with a future that they could never have dreamed about.

Our non-profit is a tax-exempt organization under section 501(c)(3) of the Internal Revenue Code in the USA. This means that any donation made to our charity will be provided with official receipts for tax deduction purposes. All of the proceeds and royalties (in perpetuity) from this book will be donated to Winning The Hopes Inc., and 100% of those funds will be used to help kids in need.

If you like this book, please consider making a donation to our non-profit and join us in the journey to affect real change. https://winningthehopes.org

CHAPTER 1

He's No Kim Kardashian, But I Still Love Him

∞

I didn't know what to expect when Shom invited me to go on a road trip with him. It was an honor for sure, so there was no way I was going to turn it down. He lived in Connecticut. I lived in Chicago. We were going to meet in Indianapolis for a series of meetings, then drive to St. Louis to host his team meeting and circle back to Chicago so he could present to my team.

My wife didn't know it at the time but I had a man crush on Shom. In the movie *Good Will Hunting*, a brilliant mathematics professor named Gerald Lambeau tries to help a self-taught genius played by Matt Damon get back on the right track from his dead-end ways.

Professor Lambeau is painfully aware he's around genius. The type of genius he knows he will never attain no matter how hard he tries. Will (played by Matt Damon) has a gift that has been bestowed on him by the gods. That's how I felt being around Shom.

Shom was a director responsible for data and analytics at the customer I was looking after. I savored every moment with him the way that a connoisseur savors a vintage bottle of Château Lafite. Not because Shom was my client and customer time trumps everything else, but because I knew I was in the presence of greatness.

Shom had a unique way of looking at situations. He could diagnose any problem and turn it on its head so that you had a completely unique perspective. Those different points of view eventually helped you make a better decision. It was first principles thinking at its best.

In our meetings in Indianapolis, Shom introduced me to some of the top-level executives in his company. As a sales rep, I always prided myself on my meticulous preparation for sales calls. I did extensive research on everyone I called on, put together a call plan, and executed with the precision of Floyd Mayweather. Shom completely shut that down.

"Bro, don't worry so much. It will be fine."

I had no option but to take Shom's word for it. We went into the meetings and I felt completely unprepared. Somehow, when it was over, it always turned out well. It wasn't until months later that I realized what had happened. While I needed pen, paper and a computer to organize my thoughts, Shom just used his gray matter. That's why I never saw him carry anything around. He just did it all in his head.

When we got to St. Louis, Shom did the same thing. I was completely amazed. He hosted a two-day planning session with his team and didn't use one PowerPoint slide or even his computer. He had two people on his team typing furiously so that they could capture every word that came out of his mouth. In that moment, I knew exactly how Professor Lambeau felt. It was like meeting Einstein or Mozart. You knew they had been touched by the hand of God and no matter how hard you tried, you weren't going to ever attain a fraction of the greatness they had.

In sales, we are taught to portray a certain image because it has been scientifically proven to work. Sales managers have a propensity to hire good-looking people. Studies have shown that attractive people are paid an average of 3-4% more than a person with below-average looks (Hamermesh, 2011). Daniel Hamermesh, professor of economics at the University of Texas at Austin and author of the book <u>Beauty Pays: Why Attractive People Are More Successful</u> has also shown how physical attractiveness is correlated with buyer impressions of likeability, trustworthiness, communication skills and a salesperson's adeptness at selling.

Another bias that benefits attractive people is called the *halo effect* (Hamermesh, 2011). This refers to the act of favorably judging an attractive person's character, in spite of any relevant facts. For example, if as a sales rep you came across as attractive and friendly, the customer might also make the false assumption that you were smart and capable (Johnstone, 2016).

As a salesperson, you are trained to always look sharp. You play up the image and wear tailored suits, drive German exotics, and have a collection of shoes that would make the protagonists in the hit show *Sex And The City* jealous. Customers see an attractive, charming sales rep and they give you a chance to pitch to them. Bingo! You've just played the halo effect to your advantage.

Here's the problem with that theory: you live and preach the lie long enough, you eventually start to believe it yourself. And when that happens, you lose perspective about what's important. That's exactly what was happening to me.

Eight months earlier.

I was looking after the largest health care provider in the USA. I was trying to sell them a big data and analytics solution. I zigged and zagged and touched as many of the executives and directors as possible. Unfortunately, they all pointed me back to the same person. They said that if I wanted to do anything in data and analytics, I would have to get Shom's buy off on it. Shom was the authoritative source of data strategy at their company.

I set up a meeting with Shom to grab a coffee with him. I was running full throttle and ready to turn on the charm. I'd talk to him about sports, take him to the best restaurants and schmooze him until he couldn't take it anymore. At the designated date and time, I picked him up at his office.

"Bro, let's just go outside and talk."

Instead of going for a coffee, Shom just wanted a smoke break. We went outside of the building to the designated smoker's corner. As we stood out there, I smiled and nodded my head in agreement to whatever Shom was saying. But inside, I was thinking, "What the frig?"

I looked at Shom in hidden disbelief. I couldn't believe everyone was raving about this guy like he was Justin Bieber or Justin Trudeau. Pleeease! Shom was of Indian descent. He was about 60 lbs. overweight. He had a moustache straight out of the show Magnum P.I. and wore clothes right off the discount rack. And when he talked, he had this weird staccato. The accent, along with the pace and cadence, varied so much that it was hard to follow.

I don't think I made an impression on Shom. I know he didn't make an impression on me. There's got to be more to this. What am I missing? I thought about it long and hard. That many people can't be wrong. Everyone that I trusted and respected said that Shom was the man. I had to give this another chance. I arranged to grab coffee with him again a couple of weeks later.

This time, we met at the Starbucks down the street. It was perfect. We sat outside so that he could smoke while I sipped my venti Pike. I asked a few questions. We both conversed and watched people come and go from the store. In that moment of caffeine-infused clarity, it finally hit me. As I was watching the people in the background, the only thing I concentrated on was what he was saying. And what he was saying was complete gold. I wasn't judging Shom by what he was wearing or how he looked. I was judging Shom by his clarity of thought. Shom was a complete intellectual savant. You could throw any problem at him regardless of discipline or field of knowledge and he could help you dissect it by looking at everything differently.

Over the next few months we were inseparable. I was the Luke to his Yoda. Shom helped me think through navigating complex relationships, as well as packaging my solution so that it would resonate with the leadership team. I was attracted to his brilliance like a moth to a flame. Over time, I got to know Shom better and liked him even more. He had a big heart, was fiercely loyal, and was always willing to help.

Being a big data guy, Shom recognized early on all of the biases and preconceived notions people have about appearances (Johnstone, 2016). While we were all zigging, he was zagging. He didn't put up a facade: he just focused on the work and let that speak for itself.

As our relationship developed, Shom wasn't a prospect or customer anymore, he became my brother. We grabbed multiple coffees every day and I followed him everywhere he went. I even started calling everyone "bro" just because that's what Shom did.

Your parents and teachers drill it into you as a kid to never judge a book by its cover. As we grow up, we forget that important lesson and start to look at things more superficially than we should. Shom grounded me and reminded me what was important in our relationships. It's what's on the inside that counts.

As a sales rep, when you are draped in Armani and driving your sportscar, how do you look at customers or the people around you? Are you judging how they look or are you seeing them for who they really are?

Lessons Learned:

- ➤ Don't judge a person by their appearances, the clothes they wear, or other superficial items. Material possessions and other trappings of success are nice but don't mean all that much. It's what's on the inside that counts.

- ➤ When you are talking to a customer, peer, or partner, focus on the important things: clarity of thought, depth of knowledge, and what people bring to the table.

- ➤ If you disagree with someone, pause and take a step back. Are you disagreeing with them because you disagree with their idea or are you disagreeing because of the person who delivered the message? And if it is the person, then ask yourself: What is it about that person that you disagree with?

- ➤ When you are in the presence of greatness or someone who is truly gifted, savor the moment. Be thankful that you have been given the chance to witness such excellence.

- ➤ Know your limitations and use techniques to compensate for them. For example, if you don't have a good memory, use pen, paper, or a computer to take copious notes.

➢ Stop comparing yourself to others. Everyone has their own unique skills and talents. Celebrate and use your gifts to their fullest potential. Others can't do what you can do.

➢ When dealing with customers, focus on the value that you can bring to them. Focus on how you can help them improve their business or lower their costs. Smart customers will focus on the message rather than the messenger. That alone can make up for any shortcomings you might have in terms of physical appearance and pizzazz.

➢ Likeability is very subjective. However, if you find a situation where there is unanimous support for one position or another, then there is probably some merit or truth to it. Do your due diligence but you will most likely end up in agreement with how most people feel about that specific person or situation.

CHAPTER 2

My "Twin" Sister – A Better Version of Me

∽∞∾

"You go to school. You be engineer or doctor. You make lots money. Make mom and dad proud. After, you marry good Vietnamese girl. She make you happy. We pick for you," my dad said with the conviction of a Buddhist monk.

"Dad, I don't want to do that. I don't want to be an engineer or doctor. I don't like those things. I want to go into business. You can make lots of money in business too."

"Bit-ness? What bit-ness? You become doctor or engineer. Those are good jobs. You be good boy and listen."

I was in grade nine and about to make the toughest decision of my life at that time. Science was a mandatory subject up to grade ten. If you were going to go become a doctor or an engineer, you had to continue taking advanced science classes right through grade twelve. I hated science more than kids hate eating their vegetables. My brain just wasn't wired that way. But I had an affinity for math because it was the root of what I was most interested in – money.

Big life events never just happen all at once. You don't wake up one day and find out that you can play professional basketball, that you are an accomplished author, or that your spouse suddenly hates you. Who you are and what happens to you are the result of millions of micro decisions made long before you ever realize the repercussions. It's years in the making.

I decided then and there that I wasn't going to listen to my parents. I had to follow my passion and go into business. I didn't know what that really meant or what I was going to do in business, but it was what I wanted to do. After grade ten, I would not take any more science classes. I was going to focus purely on math, economics, and languages. To do well in business, you had to be able to communicate effectively. I decided to focus on English and Latin because they were the basis of all business communications.

My family and I were refugees from war-torn Vietnam. We were part of the boat people that were sponsored by compassionate groups like the Catholic Church and given a new lease on life (literally). They sponsored us and gave us a second chance by relocating us to Canada. They helped get my parents jobs and made sure my sisters and I got first rate educations in all-boy and all-girl Catholic schools.

My mom worked in a plastics factory for minimum wage. My dad was a refrigeration technician. We knew we were dirt poor, but thankful for an opportunity to make a better life for ourselves. My parents were very traditional. They grew up with very strict Southeast Asian customs and mores and were going to make sure their kids inherited those same beliefs.

When you are uneducated, you always have a chip on your shoulder. You wonder if you are ever good enough or smart enough compared to those who have university degrees. What you crave the most is **respectability**. That's why my parents were pushing us to become doctors and engineers. Those were known and respected professions. In the Asian culture, when you can say that your child is an engineer or a doctor, that's like saying your son is Michael Jordan or LeBron James. It doesn't get much better than that.

When you are dirt poor, you desire an opportunity to make not only a lot of money — but **consistent** money. That's why high-paying professions like doctors and engineers are revered by poor immigrants. They are secure, safe, and a sure bet that you will have a better life.

For the next eight years, I made my parents' lives a living hell. When they said the sky was blue, I wrote them an essay as to why I thought it was red. All in a language (English) that they could barely understand. I chipped away at everything they believed in. I dove head-first into the rollercoaster that is business. I went door to door selling household products. In their eyes I became a used car salesman. That was the lowest of the low.

Out of the blue, just to stick it to them even further, I brought a white girl home and told them that I was going to marry her. We had just gotten engaged. They pleaded and begged and told me I was making the biggest mistake of my life. When they realized I wasn't going to change my mind, they didn't talk to the white girl and me for the next three months.

I remember vividly the moment everything changed. I had finished my undergraduate degree and started my first job. It was a sales job in the burgeoning field of high tech. All of a sudden I was making boatloads of money. More money than I had ever dreamed of. I started paying rent and giving them gifts. They still didn't understand what bitness was, but they knew the smell of success. Overnight I became the favorite child. The prodigal son returns. All it took was money.

Twenty years later.

My colleagues and I were at a networking event in New York City. We were having fun socializing and mingling. Benjamin Prime's is one of those classic New York steak houses. The atmosphere is festive, the food fabulous, and the staff knew our names because we went there every week.

"Are you Vietnamese? I can tell by your name."

I had just been introduced to Liz by Peter, our alliances manager. Liz worked for one of our large systems integration partners. For the next half hour, I got to know her in a deep and meaningful way. Five minutes into our conversation, I realized I had met my doppelganger (universal twin). Every experience I have had over the last twenty-five years, every emotion I have felt, and every success I have ever achieved was matched by Liz. The only difference: she was female.

"Liz, your parents must be so proud of you!" I said after my sip of Tanqueray and tonic. What's not to be proud of? Liz was highly educated, had a successful career, made lots of money and gave a lot of it back to support her parents.

"Well, actually no. I'm still the black sheep of the family."

My heart sank and broke into a million pieces. Liz had done everything that I did and she was the outcast of the family. She had bought her parents a house, a car, and provided everything that they needed. She supported them the way any loving kid would. But her parents were so traditional that they believed she should have stayed at home and been a good wife. Her role was to take care of her husband – an Asian one – and raise the kids. To them, that was the pinnacle of womanhood.

The lightbulb finally came on. All this time, I thought money is what made me the favorite child. It turns out, that was only a small part. In Asian and other very traditional cultures, it was all about my sex. I was the golden **boy**.

We have been talking about gender inequality for the last forty years. To be sure, we've made progress, but not as much progress as needed. According to the Department of Labor (2018), women make up 47% of the workforce in the U.S., and 58.6% of women age 16 or older are participating in the labor market (Hunt, Layton, & Prince, 2015). Women enter the workforce in roughly equal numbers to men, and yet they account for just 14.6% of executive officers, 8.1% of top earners, and 4.6% of Fortune 500 CEOs, according to American Progress (Hunt, Layton, & Prince, 2015).

In a joint Intel and Dalberg report, tech companies with even one female leader had a 13%–16% higher enterprise value (controlled for age, size, profitability, and revenue) than firms with all-male leadership (Dalberg Advisors & Intel, 2016). In another report from First Round Capital, their investments in companies with female founders performed 63% better than companies with all-male founding teams (First Round Capital, 2019).

And while women are graduating from college in greater numbers than men, and earning comparable salaries when they enter the workforce, many find themselves taking a step back just when their careers should be taking off. The problem arises when employees try to balance work and family and women end up carrying nearly all of the caregiving responsibilities (Dalberg Advisors & Intel, 2016).

Our current cultural support systems are innately biased towards males, but there is hope. We are starting to see companies in different industries institute women-in-leadership programs and being more flexible in the workplace to accommodate women returning to the workforce after taking time off to start families. Very progressive companies are even removing names from the resume screening process so they can select candidates for interviews based only on education, experience and other required qualifications. They are also paying people market-rate salaries versus making salary decisions based on salary history (Hunt, Layton, & Prince, 2015).

Studies have shown a correlation between companies committing themselves to diverse leadership and achieving greater success (Hunt, Layton, & Prince, 2015). The more diverse companies are, the more they are able to attract top talent, as well as improve their customer orientation, employee satisfaction, and decision making. McKinsey did an exhaustive study that came to the following conclusions about diversity and its impact on businesses:

1. Companies in the top quartile for racial and ethnic diversity are 35% more likely to have financial returns above their respective national industry medians.

2. Companies in the top quartile for gender diversity are 15% more likely to have financial returns above their respective national industry medians.

3. Companies in the bottom quartile for both gender and racial and ethnic diversity are statistically less likely to achieve above-average financial returns than the average companies in the data set (that is, bottom-quartile companies are lagging rather than merely not leading).

4. In the United States, there is a linear relationship between racial and ethnic diversity and better financial performance: for every 10% increase in

racial and ethnic diversity on the senior-executive team, earnings before interest and taxes (EBIT) rise 0.8%.

5. Racial and ethnic diversity has a stronger impact than gender diversity on financial performance in the United States, perhaps because earlier efforts to increase women's representation in the top levels of business have already yielded positive results.

6. In the United Kingdom, greater gender diversity on the senior-executive team corresponded to the highest performance uplift in the data set: for every 10% increase in gender diversity, EBIT rose by 3.5%.

7. While certain industries perform better on gender diversity and other industries on ethnic and racial diversity, no industry or company is in the top quartile for both dimensions.

8. The unequal performance of companies in the same industry and the same country implies that diversity is a competitive differentiator, shifting market share toward more diverse companies (Hunt, Layton, & Prince, 2015).

Lessons Learned:

➤ Over the last twenty years, we've made great progress in terms of race and gender equality. The key question to ask ourselves is have we made enough progress? Are we all doing enough around diversity and inclusion to create a better workplace? In your daily interactions, what can you learn from the person next to you? What can you learn about their race, their culture, and how they identify themselves? Try to empathize with them and understand their point of view even if you don't agree with it. As an experiment, try and see if you can learn something new from a different culture once a week. Over time, you will gain an appreciation for what makes different people and their cultures great.

➤ We have an opportunity to learn from both our elders and our kids. Our parents and elders can teach us from the hard life lessons that shaped them into who they are. Our kids can teach us to be curious and amazed at the wonderful things they experience every day. The two extremes can teach us to appreciate the present. The next time you are with your elders, ask them to tell you a story from their past and see what you can learn from it. When you are with kids, ask them to show you something that they like doing today and learn from it.

CHAPTER 3

I Kissed A Procurement Manager

∞

I was holding hands with my girlfriend, Michelle. As we were walking through Western University's beautiful campus, my hands went clammy. I was in my last year as an undergrad and our relationship had progressed to that awkward point, when you start to ask the hard questions. And some questions you don't ask – mainly because you are not ready for the answers: Are we going to stay together? Are we going to split up? As I got ready to speak,

I could feel my heart pounding in my chest. God, I was nervous.

"Michelle, I have to tell you something."

She squeezed my hands and looked at me with those big, bright brown eyes that could melt icebergs.

"I have an interview in two weeks with a computer company in Toronto."

I figured this would be the beginning of the end of our relationship. I was moving on. She had another year left at Western. Long distance relationships never work. Let's be honest, I wasn't mature or smart enough to realize the blessing that I had – even if it was standing right there in front of me.

"Trong, that's awesome. I'll come with you!"

Over the next two weeks I diligently prepared for the interviews. Michelle helped me refine my style, approach, and how I came across. I was getting confident. I felt good. I was Rocky Balboa running up the 72 steps to the Philadelphia Museum of Art.

I breezed through the first three interviews. They had three different sales managers poke at me around three different criteria that were important to the company. I saw the questions coming from a mile away and hit each one out of the ballpark. I was hot. I was on fire. One more interview to go and I was done.

My last interview was with Peter. Peter was the VP of Sales. He would be my ultimate boss. If I couldn't win him over, then it wouldn't matter. When I did my research, I found out that in his prior role, Peter was also the VP of HR. If he was the VP of HR, then what he specialized in was people. He would see through me in a nanosecond. Sweat beads started to form and my heart pounded in my chest. I flashed back to our preparation. What did Michelle tell me?

"When in doubt, just be yourself. If it is not good enough, then maybe it is not a company you should be working for."

Michelle was definitely the smart one in our relationship.

As Peter asked me questions about why I wanted to go into sales at the company, I decided to drop the facade and just be myself **(Authentic)**. I told him that I grew up dirt poor. I needed to make a lot of money to support my family and live the lifestyle that I wanted **(Long-Term Perspective)**. I told Peter that the business I started in University was a complete failure **(Honesty)** and I hadn't made the kind of money I needed to pay for this year's tuition. I told Peter that I was going to blow out my numbers for him because failure was not an option. I left everything on the field. If I didn't get the job, then it just wasn't meant to be. Peter told me that someone would get back to me in two weeks.

I've had the pleasure of working with many talented procurement professionals. Some of those relationships were uneventful and some of those partnerships were profoundly successful. What made the difference between them? From a sales rep perspective, what makes for a great procurement person?

I've thought about it long and hard and I want to share some of my personal thoughts on the qualities and traits I value in a procurement professional. Here goes:

1. **Integrity/Honesty:** This is definitely one of the top qualities, if not *the* top quality, I look for in a partnership or relationship. It's rare, but sometimes you run into procurement professionals who are a bit fast and loose with their approach. Sales professionals might sometimes find themselves in situations or negotiations where the procurement person they are dealing with will use them as cannon fodder for a decision that has already been made.

2. **Long-Term Perspective:** I value procurement professionals who want to take a long-term perspective. They want to build relationships and make sure that their company is getting the best deal over time. They are not there just to negotiate the best immediate deal, collect their bonus, and move on. With a short-term focus, the procurement professionals don't really care if the vendor bleeds on the deal or not. They only care about getting the best immediate deal for their company. As a result, these relationships are not sustainable for the long term.

3. **Authentic:** A lot of procurement professionals I deal with put a wall up around themselves. They hide behind a facade of professionalism. Sadly, their interactions have a hint of remoteness. I always find this unfortunate, as I am not sure it is the most productive way to deal with others. When you net it out, we are all just people. We are there to do a job (and we want to do it well), but we all have personal lives and interests that make us people. The quicker we relate to and deal with each other as people, the better relationships we will form. The better relationships we have, the more likely we are to go that extra mile to make each other successful. That is

truly the mark of a good vendor/procurement relationship.

Ten years later.

Zack Sengh is one of the best procurement professionals that I have ever dealt with. He epitomizes these characteristics. Zack was the VP of Procurement at a global conglomerate that I looked after. For a period of six months, I connected with all of the executives in the company. I launched a massive campaign to get a deal done. The marketing was a success. The customer need was evident. And the relationships required to get a deal done were finally in place.

The last step was to get through procurement. When I first met with Zack, he sat me down in his office and walked me through how he liked to work with vendors, what he was looking for in a sales rep, and how he wanted to negotiate this deal **(Integrity/Honesty)**. Zack told me that he wanted to have an enduring relationship between our two companies. He wanted to make sure that his company got a good deal, but he didn't want us to bleed for it. He wanted to make sure that we had enough margin in the deal to be successful as well **(Long-Term Perspective)**.

This was music to my ears. Any sales rep who hears this starts to smile on the inside. Perfect! This is exactly what we want as well. Unfortunately, I wasn't buying any of this. I had been to this goat rodeo before; way too many times, unfortunately. Clichés are clichés for a reason. So, I'm going to use two right now – the proof will be in the pudding. I wanted to see him walk the talk.

Over the next four months of contract negotiations, Zack was true to his word. Every time he pushed us to the brink on a legal or business issue, I stood my ground and told Zack we were on the edge. At that point, Zack backed off to make sure we wouldn't bleed. Zack was incredibly good at what he did. He extracted some unbelievable concessions for his company, but he did it with integrity, transparency, and a long-term perspective.

Throughout the process, I got to know Zack better. He talked about his wife, his hobbies, and things he did that brought him joy. When we got to the closing stage and he was ready to sign, I asked him if we could do the official signing over dinner. It was a big deal for our company, and I wanted to commemorate the moment. As the wine was flowing, we let our guards down even more. Zack told me about his son who had died of cancer **(Authentic)**. In that moment, we both shed tears and realized what was truly important.

Zack was one of the most memorable buyer/seller relationships that I have ever had. What made us get along so famously was that we were mature enough to realize we were two sides of the same coin. To the extent that he was working hard to do what was best for his company, I was doing the same for mine. We had shared interests in making sure that both of us were successful in the long run because if any of the implementation was derailed, he would need my help to fix it. If we were bleeding on the deal, then there would be no incentive and probably no mechanism for me to help him out.

When sales leaders are hiring sales reps, what characteristics do you think they should be looking for? If sales and procurement are two sides of the same coin, shouldn't we be looking for the exact same attributes in a sales rep as we do in a procurement professional? What do you think?

Lessons Learned:

➢ Sales and procurement are two sides of the same coin. The common attributes in the best of breed sales reps and procurement professionals include integrity, honesty, long-term perspective, and authenticity.

➢ Procurement professionals are there to help their companies solve a problem. They try and find the best solution and get the most value for their companies. Don't treat them as adversaries. Treat them as partners and you will be part of the solution that will help transform their business.

➢ Procurement professionals are under the same pressure as sales reps. They have to get a deal done in a certain timeframe, make sure they hit the Key Performance Indicators they have been assigned, and look out for the company's best long-term interests.

➢ Beyond their business facade lies a real person. They put on their pants one leg at a time, have families, and deal with the same problems as everyone else.

➢ After a deal is done, continue to develop and hone your relationship with the procurement professional. You never know when you might need their help in the future.

CHAPTER 4

If I Could Turn Back Time

"You meet a new customer for the first time. They want to buy your product if you do these three things that they have outlined for you. They want a stair-step function in terms of discounts that start at 10% and grow 3% with each second step. And they have operations in these four countries. You got all that?"

"Yup. I got it."

"Good. Forget everything I just said. Here's what I'm really interested in."

For the next forty-five minutes I got drilled in every different direction about my background, my experiences, and why I wanted to join Dell's elite global acquisitions team. I couldn't be sure, but I thought I was doing fairly well.

"One last question. The story I told you at the beginning. Tell me what countries the customer had their operations in, do it in reverse order, and what was their second demand?"

Bam. That's how he knew if you were good enough to join his team. The last forty-five minutes had been a complete diversion. John wanted to create a realistic sales scenario just to see if I was agile enough to filter out the noise and focus on what was important.

Everyone has a favorite boss that they loved and would do absolutely anything for. John was mine. John worked hard and played even harder. He had a beautiful wife, two gorgeous kids and made so much money from his tenure at Dell that it was almost gaudy. John was my real-life version of Tony Robbins. He painted a vision of the possible and then showed you that with hard work, you can absolutely achieve these seemingly impossible goals. Tony Robbins could have learned a few things from John.

Over the next few years, my team and I achieved everything that John said we would. To be sure, there were moments when we hated him as much as we loved him. We hated him because he dished out impossible quotas with ridiculous timeframes. We loved him because he pushed and stretched us to the brink until we achieved what we thought was impossible. And through it all, we had a lot of laughs and so much fun.

If good managers inspire you to achieve the impossible, what do bad managers do? For the most part, they inspire employees to leave. Numerous studies have shown that people don't leave companies, they leave bad bosses. Seventy percent of the factors that contribute to your happiness at work are directly related to your manager. Fifty percent of employees that quit cite their managers as the reason for leaving (Sheridan, 2015). These employees report that they don't have a good relationship with their managers. Manager/employee relationship is correlated with employee engagement.

A good proxy for the strength of the relationship between manager/employee is how comfortable an employee is approaching their manager with any type of question. The more comfortable the employee is, the more engaged they are. The more engaged they are, the better the manager/employee relationship.

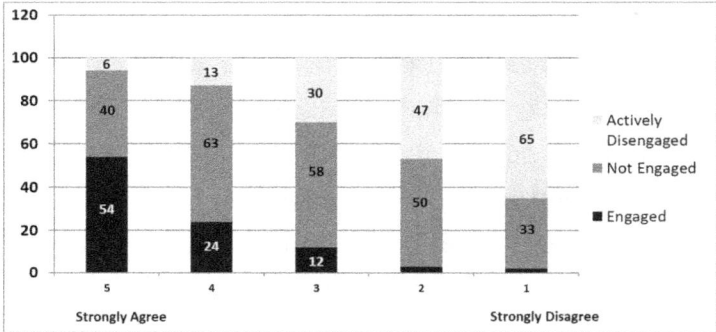

(Sheridan, 2015)

Why do employees dislike their bosses (Heathfield, 2020)? An employee could be working for a respected company with a great culture and extravagant extrinsic rewards and still quit. Kevin Sheridan with the Association for Talent Development has outlined the top nineteen traits of bad bosses. I've highlighted ten below. Can you identify any of these characteristics in the bad bosses you have had (Sheridan, 2015)?

1. **Narcissist:** It's all about them.

2. **Screamer:** They yell and scream at you.

3. **Bully:** They manage through fearmongering.

4. **Unapologetic:** It's never their fault.

5. **Suck Up:** They spend most of their time managing up.

6. **Poor Communicator:** They are as clear as mud. They guard information and treat it as power.

7. **Terrible Listener:** They don't listen.

8. **Always Right:** They feel the need to always be right and have the last word.

9. **Unavailable:** You are left to figure out everything for yourself.

10. **Never Praise or Encourage:** They are quick to criticize, but slow to praise.

Wendy was one of my more memorable bosses. To be fair, after you've had John as a boss, anyone that follows would be a huge step down. Where John was in your face and a hard charger, Wendy was timid and demure. She reminded me of someone's grandma. Wendy always asked a lot of questions, but never provided any answers. On top of that, she couldn't make a decision if her life depended on it.

Indecision drives me crazy the way your kids drive you crazy when you know they are right.

Wendy didn't exude confidence and stylistically she was my polar opposite. She was a nice person but didn't inspire you to go take the beaches of Normandy. I was young, brash and full of arrogance. I had just won some game-changing deals and had the world eating out of my hands. I was the poster child for sales success and wasn't going to listen to someone who looked like Miss Daisy.

We finally crossed the point of no return when I was having issues with my systems engineer. For one reason or another, he and I never got along. It was like oil and water. I felt that if she was a good manager, she would have done the right thing and split us up. Over a period of a few months, tempers flared, emotions got to the breaking point and words were exchanged that were hard to take back.

It got to the point where I began to completely disengage. I stopped talking to her and just started doing my own thing. I even rolled my eyes in front of her when she made what I thought was a completely ridiculous statement. I'm sure Wendy saw it too. I have no doubt that I was on track to becoming another statistic, where an employee leaves a great company because of their boss.

A few years later – with the benefit of maturity and hindsight.

As we get older, a few things happen. We realize we should have taken better care of our bodies, we should have lived in the moment more and, if we could, we would take back some of the dumb things we did in our misguided youth. Age and maturity taught me that Wendy wasn't a bad boss. If anything, she was a pretty good boss. She exhibited none of the nineteen characteristics that Kevin Sheridan talked about in his report. She tried to coach me through a difficult situation with my systems engineer but I was just too immature to understand or appreciate it. I let my ego take charge and what was a small issue turned into a big disaster. Instead of working toward peace, I fanned the burning embers even more.

Business relationships are no different than personal relationships: it takes two to tango. In this case, it was three. There may have been some small things that Wendy could have done differently but she definitely was not a bad boss. Prior to my arrival, a lot of people at the company loved her. In hindsight, my systems engineer wasn't that bad either. He had some idiosyncrasies, but don't we all? Isn't that what makes us unique and special? As for me, I wish I could go back a few years and give my younger self some much needed advice and guidance.

If you are in a situation where you hate your boss and are thinking of leaving, ask yourself. Is it really your boss or have you played a role in making a mountain out of a mole hill? Is he or she your Wendy? Instead of quickly pointing the finger, what can you do to mediate and make the situation better?

Lessons Learned:

➢ When you are in a difficult situation with your boss, try to remove the emotion and be intellectually honest with yourself. What did you do to make the situation better or worse? What could you have done differently? The next time your boss offers a suggestion and you don't agree with it, try this experiment: Pause. Do what your boss suggests and see where it takes you. You might be surprised that they are right and their suggestions are pretty good.

➢ You can't win a battle with your boss. They outrank you and have positional authority. It is akin to a child having a fight with their parents. It is a lose-lose situation.

➢ Find different ways to make your boss's job easier: take on additional responsibilities; do more than what is required; find ways for everyone on the team to win.

➢ If you do end up working for a boss that exhibits many of the characteristics above, then act quickly and find someone else to work for. Life is too short to work for a bad boss. We spend way too much time at work to dread every minute of it.

Trong Nguyen

CHAPTER 5

An Englishman In Chicago

∽

People vote with their feet and wallets – and the people have clearly voted that they love James Bond. From *Dr. No* to *Skyfall*, the longstanding movie franchise has grossed approximately $4.8 billion USD. When you think about it, what's not to like? James Bond is smooth, sophisticated and a trained killer. Men all want to be him. Women all want to be with him.

I finally met the real-life version of James Bond about six months into working for Dell. Deep down, I think I was just postponing the inevitable. As soon as any of my colleagues from Dell found out that I lived in Chicago, they all told me to go spend some time with Pat. What was eerie about the recommendation was that it was universal. Everyone said that Pat was one of the best sales reps in the company. Pat was smart, customers loved him, and he had made a fortune getting into Dell at such an early stage. My colleagues said I could learn a lot from Pat. Clearly these guys from Dell didn't know who I was. I had no doubt I could teach Pat a few things.

Curious, I arranged to grab lunch with Pat. We met at an Italian restaurant in Rosemont. I came early and did a bunch of emails on my Blackberry. God, I loved that keyboard. I could write War and Peace on that thing in under a minute. My fingers loved that keyboard more than they love a pound of bacon.

I sensed his arrival and looked up. Shit. Dammit. I hate him already.

"You must be Trong. I'm Pat. I've heard so much about you. Thank you so much for arranging this. I'm deeply grateful for having this opportunity to spend some time with you."

He was exactly how everyone described him. He was damn good looking and his fine Italian suit fit him like a glove. Clearly it was bespoke. And when he talked, I shook my head even more. He was articulate, smooth, and thoughtful. God had blessed this cretin with a silver tongue. Not a problem. I'll put him in his place. I've been around these types of guys before. They were really good looking and smooth talking, but it was just a veneer. No substance. I got this!

I asked Pat about transforming a customer's business processes and how we could go about building the right business value assessments for clients so that they can get the funding needed to support the projects we wanted to do. Without skipping a beat, Pat went into a ten-minute dialogue about Porter's Five Forces analysis and how we could build the right ROI using different regression models. That's when I became like all the other Dell folks. I fell in love with Pat, too. He had steak with that sizzle.

Every time I got together with Pat, I learned something new. He was well read and kept company with the top leadership teams in Chicago. Hence, he usually had the latest thought-provoking insights. I would always leave my meetings with Pat with the same question:

"Pat, dude, why are you just an Account Executive here? Why don't you go into leadership and really make an impact on the company? I know you would be phenomenal at it."

Pat always laughed and shrugged. Then we would head our separate ways until next time.

Working at Dell is like being a heroin or adrenalin junkie. It was intoxicating and all-consuming. I loved every minute of it. I would work sixteen-hour days, seven days a week, and be upset that I needed so much sleep. We were on a mission to populate the planet with PCs and servers and I was its chief evangelist.

I flew American Airlines so often the flight attendants at O'Hare knew me by my first name. In a period of two years, I racked up one million miles and was entitled to every perk and benefit they had to offer. I mastered the art of taking the red eye from San Francisco to Chicago, showering in the Admirals Club, and then running to the other terminal to catch my 7 am connection to New York. Shoot, George Clooney (*Up In The Air*) had nothing on me.

All my hard work started to pay off. I was closing one massive deal after another after another. I was knocking down logos faster than Donald Trump could tweet. My work was starting to a create a buzz among my peers. This only motivated me to work harder. Of course, as wise economists have noted, there is no such thing as a free lunch. There was a price to pay for my success.

On the home front, things started to slowly unravel. I would come home completely exhausted, almost comatose. I poured my heart and soul into work and had no energy left for my family. I was so exhausted that little things would completely set me off. If I didn't have the right ingredients for cooking, I had a meltdown. If there was no bacon in the house, I went on a rampage. It was grounds for divorce. How hard is it to stock the fridge with bacon? That's it. It's over!

Stress in moderation is a good thing (National Institute of Mental Health, 2019). When we experience stress, our bodies immediately release chemicals called adrenalin and cortisol. This is our "fight or flight" instinct. Your heart races, your breath quickens, and your muscles get ready for action. It's what has helped us to survive and evolve since our caveman days. Unfortunately, when this stress response gets triggered day after day, it could put your health at serious risk.

Scientists have shown that chronic stress has the following effects on people:

Cognitive	Emotional
Poor judgement	Moodiness
Inability to concentrate	Irritability
'Brain Fog'	Fatalistic thinking
Indecision	Panic
Starting many tasks but achieving little	Cynicism
	Anxiety
Self-doubt	Feeling overwhelmed
	Frustration
Physical	**Behavioral**
Rapid heartbeat	Increased intake in alcohol, cigarettes and caffeine to relax
Aches and pains	
Frequent colds	
Skin complaints	Isolating yourself from others
Indigestion	
High blood pressure	Sleeping too little or too much
	Demotivated
	Loss of sense of humor

(Stress Management Society, 2019)

A decade had passed since I first met Pat. I was back in Chicago for some business meetings and called my friends Pat and David to get together for dinner and drinks. We laughed and shared stories about the good old days at Dell. After two bottles of wine, I thought I would ask Pat what I always asked him.

"Pat, how come you never went into leadership at Dell? I always thought you were exceptional. You could have been an outstanding VP. Why did you stay a sales rep for all these years?"

Pat finally opened up and told me. A long time ago, they had asked him to be a manager at Dell. He took the opportunity. He really liked the job, but he found that the stress changed him. He didn't like the person that he became and he didn't like how it affected his marriage. Pat said that his wife meant more to him than anything else in the world. If the decision was to choose between Dell and his wife, he would make the same choice every time. That's why he gave up dreams of leadership. Dammit, I hate Pat. Good looking, smooth, suave, smart, and *always* right.

Lessons Learned:

➢ Accumulating a million miles on an airline frequent flyer program is not a badge of honor. The sacrifices made and the time away from family are not always worth it.

➢ Fatigue manifests itself in behavior that may not be acceptable. Be self-aware enough to know when you should just sleep before you respond or react to an irritating situation.

➢ There are a lot of very talented people in the world who could do anything they want. They could be CEOs of companies if they chose to. However, they make personal and life choices not to go down that route. And that's perfectly OK. That just means they value something more in life than making money or chasing a title.

➢ When you meet good people that you like, always stay close to them no matter how far apart you are physically. Stay in touch. Those relationships are worth more than bitcoin at its peak.

➢ Some folks are just blessed: they are smart, articulate, good looking and just good people. It is OK to be jealous of them. Just be grateful that you know a real-life Superman/Wonder Woman.

CHAPTER 6:

My Grass Is Brown – And So Is Yours

∽

I was only in elementary school when Oliver Stone released his masterpiece *Wall Street* in 1987. That movie forever changed the path that I would pursue. Deep down I knew it was a morality tale that warned you about the personal price you pay chasing false gods and money. But I didn't care. I still cheered for Gordon Gekko and believed that "Greed is Good!" Gordon was my hero! Bud Fox was a great guy, but his downfall was that he had a conscience. I definitely wasn't going to make that mistake. The excesses of the 1980s and what money could buy were too intoxicating.

Fast forward twenty years and I was living a more updated version of *Wall Street*. Al Gore had invented the internet and it had taken off. Wall Street was like the color pink – so passé. If you wanted to become a millionaire, you moved to New York and became a banker. If you wanted to become a billionaire, you went to Silicon Valley and worked in high tech. Your name didn't have to be Zuckerberg, Gates, or Jobs. You didn't need to start your own high-tech company. You just had to join as part of the leadership team of a tech unicorn and, overnight, you would be worth hundreds of millions of dollars.

I loved keeping up with the Joneses. It was so fun! Conspicuous would be an understatement for my level of consumption. It was a game of one-upmanship that I was not willing to lose. My friends and peers were all playing the game with me. We were trying to outdo each other and loving every moment of it.

My boss Stephanie was my hero and everything I aspired to be. She made so much money that Warren Buffet would be jealous. She had the fancy homes, fast sports cars and tony clothes fresh off the runways. This young padawan was going to learn at the foot of Stephanie.

"Trong, why are you always so tired? You're no fun anymore. All you want to do on the weekends now is sleep and crawl in your little cocoon! Where's the excitement, Trong? Where's the Trong I fell in love with? "

"Listen. You have no idea. I'm killing myself so we can have lots of money and buy everything we want. You don't appreciate the sacrifices I'm making for you and our family."

I twisted the situation around and pointed the finger right back at my wife. It's her fault. I'm always right … right?

"Trong, we were happy when we had no money. Now we have a lot of money and we are miserable. I don't need all of these things. <u>We</u> don't need all of these things. We need to re-look at what we are doing here."

With that my heart broke into a thousand pieces. In my mind, I was killing myself to support my wife and family. In the depths of my rationalization, I was killing myself to make all this money because I thought we needed the money to buy things. The more things we had, the happier we would be. We needed the best of everything if we truly wanted to be happy.

People have preconceived notions of happiness. Like me, many believe that accumulating material possessions and money will make up for everything (Barber, 2009). Unfortunately, we fall prey to clever marketing and visual images of what happiness is or ought to be. Academics and behavioral psychologists have studied the science of happiness and determined there are things everyone can do to increase their probability of happiness (Hardy, 2015):

1. **Gratitude:** The act of being thankful releases powerful chemicals in your body that make you feel happy. The practice of gratitude can increase happiness levels by around 25%.

2. **Social Connections:** Having a robust social network is linked to positive health factors, including a longer life. Studies have shown that having ten or more friends increases your level of happiness. Strong, deep and meaningful relationships matter more than casual contacts. People who regularly spend a quarter of their hours each day with family and friends are 12 times more likely to report feeling joyful rather than feeling stressed or anxious (Hardy, 2015).

3. **Experiences:** Spending money on experiences, as opposed to material objects, increases happiness. The anticipation of the experience, the experience itself, and then the memories of the experience release dopamine in your body that makes you feel happy. Studies have shown that in North America, you only need an annual income of $75K to be truly happy (Fottrell, 2018). That level of income provides for the basic necessities so that you are not anxious or worried about money. Anything more than that won't necessarily make you any happier (Fottrell, 2018).

It was 6:30 pm on a Friday evening, and Stephanie and I were at the Cincinnati airport. Our flight was delayed back to Chicago. It had been a long but productive day of meetings. This was the last leg of a four-day trip. We were exhausted and just wanted to get home to our families. Over a glass of sauvignon blanc we started to get personal.

"Stephanie, I don't know how you do it. You have everything. How do you make it look so effortless? You are successful in your career and have a wonderful family as well. I'm jealous. I'm not doing a good job of balancing family and work."

"Trong, it's not like that at all. I work really hard and I have a great career. I also have two kids in therapy and they can't understand why mommy is never home. And to be honest, I don't know how to answer them. That's something that I have to live with."

Lessons Learned:

➤ The grass is not always greener on the other side. We just think it is. You can increase your probability of being happy by practicing gratitude, having deep and meaningful relationships, and spending money on experiences rather than material possessions.

➤ Eating healthy, exercising, and expanding your mind by reading and learning can all increase your level of happiness. The next time you are stressed, do one of these things and see how it makes you feel – work out, practice being thankful, or think happy thoughts.

➤ Family is the most important thing. It is more important than your job, your possessions, and the small issues that cause you stress on a daily basis. The next time you are stressed out, try and keep things into perspective.

➤ Keeping up with the Joneses is a fool's errand. It will never end and is also a waste of money. You only need a certain amount of money to be comfortable and happy. Anything after that amount won't make you any happier.

Trong Nguyen

CHAPTER 7

Oh, Oh, Trouble, Trouble, Trouble

∞

11:58 pm. We were at a dance club in Austin, TX. The DJ was rocking some sick beats. Everyone was drinking a bit more than they should have. We were on the dance floor, making complete fools of ourselves. We didn't care because we were having fun. We were all working for Dell and it was one of our usual monthly team meetings. Our whole team was there – Kandy, Ric, Nancy Drew, me and, of course, Jay C. No one forgets Jay. Taylor Swift wasn't born yet, but if she was, she'd write a hit song about Jay that went like this:

Oh, oh, trouble, trouble, trouble
Oh, oh, trouble, trouble, trouble
I knew you were trouble when you walked in...

The first time I met Jay, I liked him right away. He's got an infectious smile and personality. He draws people in the way a magnet attracts metal objects. It just happens naturally. Jay was 6' 3", athletically built, and always stylishly dressed. Guys gravitated to him because he was just one of the guys. He didn't put on any pretenses. He could geek out with you on video games just as easily as he could shoot hoops. And the females were attracted to him because he was so damn good looking. What got the ladies every time was his eyes. There was always a mischievous spark in his eyes that told you that he was trouble. We knew this instinctively, but we always came back for more.

Every time we got together in the same city, we would go out and have a blast. Jay knew the hot spot to go to on any given day. He knew all the bouncers, waitresses and bartenders. They treated him like family because he was. I always bailed just before midnight because – as your mother told you – nothing good happens after 12 am. The next morning, I would hear all the stories about Jay and the gang crawling into bed at daybreak.

Dell Quarterly Business Reviews are one of the most intense sales inspections anyone will go through. It is 60 minutes of hell that you wouldn't wish upon your worst enemies. You get in a room with all of your peers and the whole management team. You present your business plan for the quarter and how you are going to grow your business while keeping the margins at an acceptable level. Everyone in the room has full autonomy (and are encouraged) to tear you to pieces and pick any hole they see in your game plan or thought processes. Anyone who messes up at the QBRs usually doesn't last long at Dell. The leadership team would weed you out a few months after that.

Our whole team was relatively new so this was our first time going through the exercise. We had all prepared for a good week in advance. Our manager prepped us for what to expect, drilled into us what were the most important aspects of the QBR, and begged us to make sure we got a lot of rest the night before because we all needed to be super sharp. It was Super Bowl Sunday and you had to have your A game on.

The night before the QBR, we had dinner at a Mexican restaurant near our hotel. At 10 pm, we decided to call it a night and go back to our rooms to review and rehearse for the last time. It felt like high school or college finals: we were cramming until the very last minute. At 7:30 am the next morning, we strolled into the Dell conference room, fresh and perky from an early night. We were ready to do this thing!

Meetings at Dell always start on time. If you are not there, then that's your problem. We had an 8 am start and would continue until 6 pm. Then we would have dinner and drinks with the whole team. At 7:55 am, Jay strolled into the room. Apparently, Jay had been out with friends until 5 am, went back to his hotel, showered and *then* started preparing for the QBR. He walked around, shook hands and greeted everyone. When it was his turn to present, Jay got up and completely nailed it. Drop the mic! He played the room like a master conductor. When he finished I was in complete disbelief.

"How the frig did he do that?"

We all crammed and worked our butts off to get through the QBR with just a handful of bruises. Jay didn't even start preparing until two hours before.

Why do we make the choices that we do? Even if we know what the right answer is, why do we still make the wrong decision? Why do we eat or drink excessively when we know it is not good for us? Why do we not exercise regularly when we know it would provide a better quality of life for us?

The easy answer is that we are hooked on dopamine (Julson, 2018). Dopamine is a neurotransmitter that helps control the brain's reward and pleasure centers. Dopamine also helps regulate movement and emotional responses, and it enables us not only to see rewards, but to take action to move toward them. Drugs, alcohol, social interactions, and positive responses from Facebook all increase the level of dopamine in our systems. That's why we can't help but check Facebook a hundred times a day – we need the dopamine kick (Parkin, 2018). I didn't realize it at the time, but Jay is the real-life incarnation of dopamine. That's why we can't get enough of him.

Quarter end and year end are interesting times to be at sales organizations. You get to witness a spectrum of behaviors and results. As a sales rep or sales manager, if you have done your job well, it is a walk in the park. By now you've sold the customer on your solution and are in the process of wrapping up minor details, like paperwork and administration. It is operational maintenance.

If you haven't achieved your quotas, then it leads to some interesting behaviors. That's when sales managers put the squeeze on their reps. They start thinking about doing audits, unsolicited proposals, and unnatural things like giving massive discounts to pull in deals. To the extreme, some sales reps and managers start to do unethical things, like making side agreements with business partners or customers. I've been a fly on the wall and witnessed all of these behaviors. The one thing I've noticed is that, in the long run, none of these behaviors have turned out to be good decisions. Why did very capable and smart business people make such bad business decisions (Erwin, 2019)?

Harvard Business Review has done studies around bad decision making and concluded that most people who make bad decisions do so because of the following reasons:

1. **Laziness:** You didn't want to go out of your way to check facts, take initiative, or gather additional input. You used past experiences to extrapolate future results.

2. **Not anticipating unexpected events:** You assume that the worst will never happen to you, so

you don't plan for it. You won't see a market crash, get a divorce, or into any accident. You only think of the positive.

3. **Indecisiveness:** In the face of complex and changing data, you get overwhelmed and don't make a decision. You are paralyzed by the fear of making the wrong decision.

4. **Remaining locked in the past:** You continue to use the same old data and processes to make decisions. What's worked for you in the past should work for you in the future, so you see no need to change the way you do things.

5. **Having no strategic alignment:** You fail to connect the problem to the overall strategy. In the absence of a clear strategy to provide context, many solutions appear to make sense. When tightly linked to a clear strategy, the better solutions quickly begin to rise to the top.

6. **Over-dependence:** Some decisions are never made because one person is waiting for another, who in turn is waiting for someone else's decision or input.

7. **Isolation:** You don't involve others with the relevant knowledge, experience, and expertise to improve the quality of the decision. You make your decisions in isolation.

8. **Lack of technical depth:** You don't have the technical depth to address the problem. Or you don't tie the technical depth to the strategic vision. Organizations today are very complex, and even the best leaders do not have enough technical expertise to fully understand multifaceted issues.

9. **Failure to communicate the what, where, when, and how associated with their decisions:** Some good decisions become bad decisions because people don't understand – or even know about – them. Communicating a decision, its rationale and implications, is critical to the successful implementation of a decision.

(Zenger & Folkman, 2014)

Ten years later.

I just joined this rocketing cloud company. It's small, intimate, and you know everyone; but large enough that it is no longer a start-up and has enough cash in the bank to go the distance. The people and culture are mesmerizing. The dopamine is kicking in overdrive as I'm learning a new part of technology that I've never had exposure to before. I'm strengthening muscles I never even knew I had.

But I wasn't the only one. After a decade apart, Jay and I end up working for the same company at the same time. I shouldn't be surprised but I chuckle every time I think about it. It's not surprising because we were both hungry and wanted to work for a high reward high-tech company in hypergrowth. And this company was going only one way: vertical.

"Trong, let's go out. Tao is hot tonight." It was exactly how you see it in the movies: you look at the sky and see the bat signal. The phone rings, and you know what to do. You've practiced the move a thousand times. You turn on Spotify and blast Rihanna to get you in the mood. You spritz on some cologne, put on a tight t-shirt, and check the mirror to make sure whatever remaining hair you have stays in just the right place. All good. Let's go do this thing.

It was so fun and just as I remembered it. However, after a couple weeks out with Jay, my body couldn't take it anymore. We were carousing and enjoying the city like we were in our twenties. A decade had passed and nothing had changed with Jay. He still had the same youthful energy, infectious personality, and the hookup to the best bars and clubs in the city. We hung out with some friends (Bobby D and 4K) and behaved like we were brothers in grade school. The laughs flowed faster than the Tito's.

One morning I woke up and looked in the mirror and wasn't happy with what I saw. I was feeling sluggish, my brain wasn't as sharp as it should be at 6 am, and I had bags under my eyes. Seriously, when was the last time you saw an Asian with bags under their eyes? We could be 100 years old and still look like we are thirty. It was then that I knew I was starting to make some bad choices. I understood the theory behind bad decision making, but that didn't stop me. We know eating too much bacon is not good for us, so why do we fry three pounds of it every weekend (or is that just me)?

Lessons Learned:

- ➤ Don't compare yourself to others. Some people have unique gifts and talents that make them very efficient at getting things done. Focus on what needs to get done and work your butt off to do it. There is no such thing as a short cut.

- ➤ All of us did a lot of fun things in our twenties. Don't try to relive your glory days in your thirties and forties.

- ➤ Quarterly business reviews are hard. Always prepare and do the necessary work to nail them. Your management team is reviewing and judging you in these sessions, so bring your A game.

- ➤ Try to be early for every business meeting. Being on time is a rookie move. Being late is a disaster waiting to happen.

- ➤ When going into a big meeting, act like a politician running for election. Go around, shake hands and kiss babies. It will set the tone for the rest of the meeting.

- ➤ If you have a friend like Jay C, make sure you perfect the art of the Irish exit. No one wants bags under their eyes all the time.

Trong Nguyen

CHAPTER 8

Honestly – I Was Never Young And Stupid!

∞

"Stacey, I don't understand. Why can't you work later tonight?"

"Trong, it's not that I don't want to work later tonight. I just need some sleep. I need to look at this document with a fresh set of eyes."

We were on day seven of our marathon RFP sessions. I was in my twenties, full of piss and vinegar and ready to conquer the world. What I lacked in maturity, I tried to make up for in effort. Misguided youth, I know. But that was my mindset at the time.

Stacey and I were working sixteen hours a day preparing our RFP response for a global pharmaceutical company we were chasing. An opportunity like this comes every five or six years, and I was determined to make sure we didn't waste it.

All of the stars were lining up: I had spent the last eight months building the right executive relationships; I got us a seat at the table; and our client was listening. Hewlett Packard was the incumbent and had been for the last decade. Little did they know, their world was about to be turned upside down. That's assuming I could get our RFP response done right.

For as long as I can remember, I didn't need that much sleep. Four hours max and I was good to go. I could juggle a hundred things at the same time, go for weeks without sleep, and my brain was still razor sharp. Why couldn't everyone do this? It was easy.

It was Friday night at 11:30 pm. While everyone was out partying and having fun, I was turning the screws on Stacey. Stacey was my proposal manager, dedicated to helping me get this RFP response completed so that we could win this new piece of business. I could tell I was pushing her to the brink. I didn't care if we had to work twenty-six hours a day for the next two weeks. We were going to get this done and we were going to do it right. Oorah!

"Stacey, what's that? Are you crying? You can't cry. This is work. If you want to cry, do it on your own time."

Dammit. I pushed her over the edge. She was starting to whimper and sob on the phone. I didn't care. I was going to escalate this and get it fixed. If Stacey couldn't hack it, she should find another job somewhere else.

Recently, we've had a flurry of news activity regarding men behaving badly. Over thirty women have accused Harvey Weinstein of behaving inappropriately. Kevin Spacey was kicked off his hit show *House of Cards* because of inappropriate behavior. Superstar newscaster and best-selling author Bill O'Reilly had his contract terminated with Fox News because the cloud surrounding his behavior was bad for business.

Why are we so outraged by this bad behavior? Aren't we surrounded by it? Shouldn't we be used to this by now? Yes and no. The reason we are so outraged is because the behavior of these men was so egregious. The root of it is the imbalance of power. All of these men had an inordinate amount of power and they abused their position for their own self-interest.

There are three types of power: positional authority, relational authority, and expertise power (Consultants Mind, 2013). Positional authority comes from your rank, title, and status. It's the hierarchy in an organization. Relational authority comes from the trust and respect of others. Relational authority cannot be commanded and must be given freely by followers. Expertise power is when you are an expert at something. This type of power comes from a combination of your education and experience in a specific area.

Leadership effectiveness of a person with positional authority but no relational authority is limited. The opposite is true as well: leadership effectiveness of relational authority without positional authority is also limited. Studies have found that people give their all for leaders with both positional and relational authority.

Stacey didn't like me much after that episode. I didn't really care. I got the results that I wanted. We handed back a grade A RFP response. From there, I lead our teams in orals presentations and we nailed every one of them. When we were awarded the business, everyone cheered! The contract was for $50M net new business over the next three years.

Word quickly spread that I had the Midas touch. If you wanted to win and make lots of money, you wanted to be on Trong's team. I had the academic theory, the strategy, and the street smarts to apply it to real world settings. I was the new wave of leaders that was going to change the world.

I was like Karl "The Mailman" Malone: I delivered. Quarter after quarter, I blew out my numbers. I was making money faster than I could spend it. I was knocking one customer acquisition after another. I could write a book on this!

"Trong, have you thought about doing one of our 360-degree surveys? It's a great way to get anonymous feedback on what your team thinks of you and how you can grow."

I was out having dinner with one of my good friends when she suggested I do this company survey. She had found it immensely helpful in her development. Peers, colleagues, extended teammates, and managers would provide feedback in terms of what they thought of you and how you could improve. It was anonymous and the results were collated in such a way that it would provide meaningful feedback, but you couldn't extrapolate it to see who provided the responses.

This was perfect. I love having my ego stroked. There is nothing better than people reaffirming what you already knew to be true. Bring it on! I love people telling me how good I am.

One month later.

"A gifted salesman, Trong has a lot to learn about teaming."

"Trong moves at lightning speed. This is a good thing. But more often than not, he should slow down and bring his team along with him."

"Trong has a very condescending attitude and behavior for anyone he thinks is not smart or good enough to be part of his team. It is unfortunate that Trong is not mature enough to realize that different people have different skills and strengths."

I re-read the verbatim responses a hundred times. No matter how much I willed it, it wouldn't change. It was there in black and white. I was crushed. The worst thing about looking in the mirror is the reflection you might find. The thing that hurt the most was that, if I was being honest, the feedback was pretty accurate. There weren't any vindictive messages, hidden agendas, or malicious arrows. It was my team trying to help me become a better version of myself.

Lessons Learned:

➢ In big companies, you can't do everything by yourself. You need the support of your extended team to do great things. Leverage all the resources around you, and do it respectfully.

➢ Be cognizant of how you treat people. There is no excuse for being a jerk. You can get things done well, move at lightning speed and still be nice.

➢ 360-degree surveys can be very useful in helping to remove blind spots. Be open to receiving feedback and treat it as a gift. It will make you much better in the long run.

➢ Men in general have an over-estimated sense of self. We attribute success more to our abilities than we should. There are a lot of external forces that contribute to our successes. When we mature and grow into our own skin, we realize this fact pretty quickly.

➢ "I look back on the way I was then – a young, stupid kid. I want to talk to him. I want to try and talk some sense to him, tell him the way things are. But I can't." **Red, *The Shawshank Redemption***

CHAPTER 9

Billy The Kid Is A Killer!

∞

I had narrowed my search to three companies that I would want to go work for. All of them were born in the cloud, software as a service (SaaS) tech unicorns. A high-tech unicorn is a privately held startup company valued at over $1 billion. The term was coined in 2013 by venture capitalist Aileen Lee, choosing the mythical animal to represent the statistical rarity of such successful ventures (Zetlin, 2018). Since that time, the term has been widely adopted in our daily lexicon. Technically, none of the companies I was looking at were tech unicorns because they had already gone public. The one trait they all had in common was that they were tech highfliers and Wall Street darlings in the software space.

The company I was most interested in focused on a niche area of technology called IT Service Management. I had no idea what IT Service Management meant until I started doing research. In a nutshell, this company had software that would automate mundane tasks in large companies, such as password resets, provisioning of IT products, and other self-help capabilities. This automation helped save client companies millions of dollars every year, which is why this particular company took off like a rocket. The more I dug into it, the more I liked it. The technology was solid and the leadership team were proven veterans.

I had some check-the-box preliminary interviews, which I passed with flying colors. I reached out to my network and asked everyone I knew what the company was really like. They came back with one consistent answer: if you want to find out what it takes to be successful at the company, go talk to Billy.

Billy joined the company nine years earlier when it was barely a pimple on an elephant's butt. He helped build the company from the ground up, and was arguably one of the company's most successful sales reps. I also found out that Billy had just closed the largest deal in company history. He was about to get a humongous commission check – the only question left was how many commas were going to be on it. For the next two weeks, I hounded Billy until he agreed to meet up with me for drinks.

As I walked into the swanky bar, I liked it immediately. It was more Wall Street than Main Street and had an infectious upbeat vibe. Happy hour was bustling, and everyone was having a good time. Billy strolled in, impeccably dressed. He wore Zegna with the grace of someone born into money. Norm had Cheers and Billy had the Cactus Club: here, everybody knew his name. Billy didn't even have to walk around. People just came to him to pay homage and give him hugs as if he was a long-lost relative.

We settled into our first couple of drinks and I started picking Billy's brain about the company. I asked about the company culture, the people, the operational aspects, and day-to-day life there. Billy was transparent and honest. He didn't hold anything back. I found out everything I needed to from a business perspective. At this point, I decided to pivot and ask him personal questions to see if he walked the talk.

"Billy, what are you doing this weekend for fun? How are you going to chill out?"

"Trong, I don't think there is going to be much chilling out this weekend. I'm in the middle of renovating my house. I'm putting in this professional hi-fi home theater system. You would totally love it. The components alone cost me $30K. And I think my wife and I have to go to this art exhibit fundraising event on Saturday night."

Alright, I see how he rolls. Let's step it up a bit.

"What's your dream car? C4S or 911 Turbo?"

"I have a Porsche 911 GT3 RS. It's my fourth Porsche."

I had to tap out. I had nothing left. If you didn't already know this, Porsche 911's have been my dream car since I was a little kid. It's the iconic sports car that every kid dreams about. The closest I ever got to one was the poster on my bedroom wall. As an adult, I upgraded and bought the miniature replica from Amazon. Billy had a Porsche 911 GT3 RS and raced it on the weekend with friends. That's a $180K car that he drives for fun. I was completely smitten. I decided to put him through one last test.

"Billy, be honest, I heard about the massive deal you just closed. I know you blew out your numbers. Are you going to make over $1M this year?"

Billy just shrugged, smiled, and gave me a non-committal wink. Now I knew why everyone loved Billy.

Ninety percent of startups fail (Krommenhoek, 2018). That's a fact, yet every year people swarm to startups. Whether it is a calling that they couldn't ignore or they just got tired of the big corporate machine, people are going to startups hoping to strike personal or professional gold. Lured by the fortunes made in Silicon Valley, a big portion of business school students today would rather take a chance on a small startup than the guaranteed payday of Wall Street or high-end consulting companies. Most people who join startups list these reasons as their motivators for ditching the corporate world (Kan, 2018):

1. **Personal/Professional Growth:** Startups are limited in money and resources. They can't afford to hire all of the people they would like, so everyone plays multiple roles. The flip side of this is that it gives you the opportunity to wear multiple hats and make a real impact. You will get an opportunity to do things and make decisions that you wouldn't get until much later in your career in

the traditional big corporate company. While many people who leave large companies may sacrifice bigger salaries, the experience they gain at a startup is priceless.

2. **Financial Upside:** Big companies pay more in cash and upfront benefits because they can afford to. When you are working for a startup, you are essentially betting on the company and yourself. You are betting against the odds that you will succeed. When you do, the equity part of the equation will more than make up for the risk. If you get in early enough, the upside can be life changing.

3. **Culture:** In an established company, the culture is already defined. You may or may not fit into it. If it's not a good fit, then you keep quiet, grin and bear it. It becomes a job where you do your work and collect your paycheck. In a startup you get to influence and create the culture. You can use that creative spark to instill what makes you and your colleagues unique and have that as part of the DNA of the company. This is not to say that you can't do this in a larger company – it is just much, much harder.

4. **Impact:** In a startup there are usually fewer levels between you and the CEO. Fewer levels mean more opportunities to interact with the executive team. If you are really good at what you do, it will get noticed quickly and more opportunities will open up to you (Sheldon, 2012). Of course, the flip side of this is true as well: if you are not very good at what you do, people will notice very quickly. There is no place to hide in a startup.

"Billy, be honest. How were you so smart? How did you know this company was going to kill it? What made you take a bet on this company? I think I'm a pretty smart guy but I don't think I'm smart enough to pick the right startup to go to."

Billy gave me the most honest answer I've ever heard.

"Trong, I had no idea. It was pure luck. I didn't know this company would be such a massive success. I had a hunch that they were on to something, but I had no idea. I just made a bet and it paid out. Here's what happened ten years ago. I had started my own company and I needed to raise some funding so I flew down to meet all of these VCs in the Valley. As I'm sitting in the waiting area, I overheard the guy in the next room. The VCs were throwing themselves at him. His company was still unprofitable, but they were willing to give him any amount of money he needed. He turned them down and told them that he didn't need any money.

After he left, I found out who he was and the company he started. At that point, they were just starting to expand and wanted to go into the enterprise sector. I pretty much hounded him for a year to give me a job as their first enterprise sales rep. I didn't fit the profile they were looking for, but he decided to take a chance on me. We took a chance on each other and it paid off. That was it. I didn't have a crystal ball."

Lessons Learned:

- ➤ You can make a calculated bet on startups, but no one really knows who is going to be the next Amazon, Facebook or Google. It has a lot to do with luck and being at the right place at the right time.

- ➤ Being prepared with the right skills allows you to take advantage of opportunities when they serendipitously appear.

- ➤ There are upsides and downsides to going to startups and smaller companies. The upside is that you get a chance to build something from the ground up. The downside is that there is a lot of risk, minimal resources, and minimal support. You have to determine if the rewards far outweigh the risks.

- ➤ Companies are like life partners: There are a lot of great ones out there; the key is finding the right one for you. Do you like the culture, the team, and the products you will be selling? If the answer is NO to any one of these questions, then maybe the company isn't right for you.

- ➤ If you are joining a startup and the perks and benefits far outweigh those of mature companies, leave immediately. If the executive team is lavish and spends money like it is going out of style, sell the stock immediately. Smaller startups do not have the resources to roll like that. Look at all the dot.com companies that went bust during the internet boom as empirical evidence.

➢ Having money and spending it on what you want to enjoy life is nice, but it is also smart to start saving as much as you can at an early age. The power of compounding will make you rich in the blink of an eye.

CHAPTER 10

Don't Tell The Truth

∽∞∾

"Nooooooo!" I screamed as I fell off the building. Desperate to cling to anything, I reached, grabbed and clawed at air that slipped through my fingers like grains in an hourglass. I woke up in a puddle of sweat and looked at the clock. 3:51 am. Shoot. The first digit hadn't even turned four yet. I was going to drag my butt again today. Unfortunately, it was the 15th day in a row. I had been through this before. I knew what the signals meant but was not ready to face the truth. Deep down, I didn't want to break up.

Your professional life is no different than your personal life. They follow similar patterns. It goes something like this: you meet someone you like, you analyze the situation, determine if it is a good fit, and then make a commitment.

But what do you do when reality doesn't meet expectations? What happens when you feel you were oversold? Regret? Buyer's remorse? At that point, you have to determine if the benefits outweigh the baggage. I didn't want to break up, but I knew it was the right thing to do. I was on the shore and ready to cross the Rubicon. Mentally and emotionally, once you flip that switch, there is no going back.

Over the next two months, I got everything ready. I updated my resume. I talked to my peers, friends and customers to get their insights on the hottest tech companies in the industry. I researched those companies and scrutinized them more than a stuffed person in a buffet line. I narrowed it down to three possibilities. All three fit the bill in terms of culture, pay and market relevance. It was show time.

I nailed my first round of interviews with all three companies. The HR screening process is relatively easy. It's a quick analysis to see how many boxes they can tick off the laundry list of attributes they are looking for in that particular role. Also, and more importantly, they gauge to see if you are a cultural fit. Done.

The next few rounds included sessions with the hiring manager, his peers, and the hiring manager's boss. This is the fun part. They start dissecting you and see what you are made of. The good interviewers will drill into how you behave in certain situations. They will peel back the onion layer by layer and, if they do it right, it will be apparent quickly if you actually have done all the things you said in the interview. The bad interviewers (and the managers you don't want to work for) are the ones who follow a script and ask you the twenty questions they have read straight out of the HR manual.

I was two thirds of the way through the interview process with all three companies. I was going to have to narrow down my choices. The last rounds of interviews are always the toughest. Most companies today run a finals round where you get grilled by a committee of people and do a formal presentation. If you nail the finals round, then odds are high that they will make you an offer. This is definitely my favorite part of the process. It's crunch time. High stress, high pressure and you are put under a microscope. Correct that – they are doing DNA testing on you. 23andMe. Sign me up!

A bad boss is the number one reason employees quit their jobs (Klotz & Bolino, 2019). The cliché that people leave their bosses and not the companies rings true across multiple dimensions. Employees who rate their supervisor's performance poorly are four times as likely to be job hunting. If we unpack that a little more and ask why employees are so unhappy, patterns begin to emerge.

It turns out many of us have unanswered callings at work. We have passions that we did not get a chance to pursue in our careers. We might have lacked the talent, the opportunity or the means to make them occupations. We spend so much time at work that we don't have time to pursue these passions as a hobby. At that point we try and bring some of these passions into our jobs. A good manager plays a major role in motivating, providing constructive feedback, and directing the work that needs to get accomplished. Too often though, managers don't know what motivates their employees or don't have the right relationship with their subordinates. We all want to do meaningful work. Managers should help employees connect why their work has meaning or they will find a job with an employer who can.

Other major reasons why employees leave their jobs:

10. **Lack of employee recognition:** Something as simple (and free) as showing appreciation for employee contributions can be a difference maker. Nearly 22% of workers who don't feel recognized when they do great work have interviewed for a job in last three months, compared to just 12% who do feel recognized (Klotz & Bolino, 2019).

11. **Company culture is not a priority:** Employees who rate their culture poorly are 24% more likely to leave. Research found that culture has an even bigger impact on an employee's decision to stay or go than their benefits package (Schwantes, 2018).

12. **No growth opportunities:** Employees who feel they are progressing in their career are 20% more likely to stay at their companies in one year's time (Schwantes, 2018). Employees who don't feel supported in their professional goals are three times more likely to be looking for a new job.

(Kununu, 2018)

Finals day had arrived. This data company I had been talking to had a unique culture and a leadership team that I could resonate with. I was pumped. This was going to be a four-hour ordeal. They poke, they prod and they tear you to shreds. The best part is that you get to do it back to them. I was told ahead of time that after the customer pitch, they will caucus for 15 minutes. You stew in a waiting room until they come and get you. The first three hours flew by in a blink and I nailed it. The customer pitch was as easy as eating bacon. I was on fire. A few minor stumbles but overall I felt good. The hiring manager brought me to the waiting room and told me that he would be back in 10-15 minutes to get me.

Fifteen minutes goes by. Then twenty. Then thirty. I knew something was wrong. If it was an easy decision, they would have come back fairly quickly. Thirty-seven minutes later the hiring manager comes back and gets me. I knew this wasn't good. Somewhere along the way I had made a misstep or I hadn't nailed it the way that I thought I had. Either way, I wasn't going to get this job. Oh well, you can't win them all.

They sat me down and lobbed up some softballs. The company asked me what I thought of the process? They wondered what would I have done differently? They told me they liked seeing me in action and presenting. Then they dropped the bomb. Their Global VP of Sales, who for the most part had been silent and observing, asks, "We've talked to your last two managers at your current company. What do you think they would say about you? What do you think of them?"

Shoot. The jig was up. In that split-second I could feel sweat start to form on my forehead, my mouth started to get as dry as the Sahara dessert. If it was possible for an Asian, I would have turned Casper white. The rule of thumb when you are looking for a new job is to NEVER disparage your former employers or bosses. If I lied and told them how great everything was, then I would be disingenuous. If I told the truth, then at a minimum I would sound petty and self-absorbed. I extended my pregnant pause for what seemed like a full nine months. In tough situations like these I always think about what my best friend Michelle would do. Her advice hasn't changed in twenty years – **when in doubt, just be honest and tell the truth.**

"I loved my first boss. He was a great leader. Let me tell you what I liked about him …. Unfortunately, I didn't like my second boss as much. I think he was more sizzle than steak. Let me give you a few examples of what I mean …."

95

With those final words, I knew I had sealed my fate. I had broken the cardinal rule of interviewing and spoken negatively about a former boss and employer. I was ready to pack up my bags, lick my wounds, and then curse myself for being so dumb. No one is ever that honest in an interview.

After hearing my answer, the Global VP of Sales piped up. "It's interesting you said that about your old boss. That was the same reason I fired him as a manager eight years ago when he worked for me. We'd like to make you an offer to join our company."

Lessons Learned:

- ➤ When you are job hunting, do meticulous research on your potential employer, what they value, the people you will be interviewing with and what they are looking for. Be over prepared. Rehearse your answers. Practice until you can be succinct and clear on the messages that you want to get across.

- ➤ Interviews are a two-way conversation: you are interviewing them just as much as they are interviewing you. Make sure you find a company with the right fit, culture and products you want to sell. During the interview, just be yourself. If they don't like it, then it is not the right company for you. When they ask really tough questions and you don't know how to answer them, try this technique – just be honest.

- ➤ Company culture really matters. Find one that aligns to who you are as a person and what you believe in. When you are with the interviewer, ask them questions to see if they live up to the values they are talking about. If they don't, then my suggestion would be to find someone else to work for. You want to work for people who walk the talk.

Trong Nguyen

CHAPTER 11

Catching The Next Wave

When I'm with you baby
I go out of my head
And I just can't get enough
And I just can't get enough

The classic Depeche Mode song was constantly playing in my head. I was in the midst of choosing the company I would call home for the next few years, and the experience felt like I was shopping for a new home (no pun intended) again. Should we get a detached, semi-detached, townhouse or condo? I've always loved house shopping because you get to see so many different styles and ideas that you can eventually incorporate into your new purchase.

For decades the homes I've lived in were good metaphors for the companies I worked for. They were big, provided you with experiences and skills that you needed, and were necessary for that point in your life.

In a big company, I took advantage of their size and scale. I hopped around different divisions and learned what business was all about from different angles. I took every training class that was available in an attempt to improve my sales and business acumen. Life in a big company was perfect for the stage of life I was at.

Fast forward twenty-five years and now I was at a crossroads. Just like most empty nesters after their kids have left home to start their own lives, I was wondering if I really needed this big house. Somehow it didn't feel comfortable anymore. It felt way too big and had more space than we really needed to be happy. The big house was hard to maintain. It was way too much unnecessary work for the enjoyment we were getting in return.

I really couldn't complain about the big companies. They paid well, had lavish benefits and it was always a big ego boost when I told my friends and family that I worked for XYZ Corporation. Unfortunately, those bragging rights came at a cost – lots of cleaning and yard work for only two people. After two decades of working, I decided to take a different path. For the first time in my life, I would work for a small company. I decided it was the right time to move into a condo. It was smaller, didn't have all of the benefits of detached home, but it was the right fit for this time in our lives.

After a lot of soul searching, I accepted an offer to join a small cloud company that provided planning software. This space was new to me, but I knew I was smart enough to pick it up quickly. I would just have to study a bit more during my onboarding.

On my first day, they shipped me a Mac laptop. This was not going to be good. I've been a PC guy my whole life. Shoot, I even worked for the companies that arguably created the PC (IBM and Microsoft). The only thing I knew about Apple was that Steve Jobs was a maniac! I attended the first-day training with a remote group of 20 new hires. They walked us through the instructions of how to set up our Mac. Unfortunately, none of the instructions worked. It turns out that whoever imaged the new laptops didn't do a good job. None of our credentials were set up. On top of that, the company didn't even have a help desk. **Strike one**. Did I just make a big mistake? Suck it up Trong. Don't make rash judgements.

A couple of days later, I was all set up. I made an appointment to meet with my new manager in our office. It's your typical new hire meeting where we go through administrative details, set expectations and agree on a process to optimize our mutual performances. I arrive at our Grand Central Station office and notice that there are people hosting conference calls in the hall. Interesting. As I open the door, I completely understand why: the office was smaller than a can of sardines. I scanned the room and did my Rain Man trick. One, five, eight, twenty. There were twenty desks and 32 people in the office. There was actually more space in the hallways. **Strike two.**

The meeting made it clear: no corporate credit card, no more catered lunches and no more lavish customer dinners at Nobu. I hoped I wasn't at strike three yet, but I couldn't be sure. It definitely felt like it. Did I just make a bad decision?

Harvard Business Review did a comprehensive study that talked about the bad habits that lead to poor decision making. The interesting part is that these habits can easily apply to your personal life as well:

1. **Laziness:** This is a failure to check facts, to take the initiative, to confirm assumptions or to gather additional input. These people rely on past experience and expect results to be an extrapolation of the past.

2. **Not anticipating unexpected events**: People don't take the time to consider all the negative things that could happen. Too often people get so excited about a decision they have made that they never take the time to do simple due diligence.

3. **Indecisiveness**: These people are paralyzed by fear of making the wrong decisions, or believe that one mistake will ruin their careers, so they avoid the risk by not making any decision at all.

4. **Remaining locked in the past**: These people get used to approaches that worked in the past and tend not to look for new approaches that will work better. Better the devil they know than the one they don't.

5. **Having no strategic alignment:** Bad decisions sometimes stem from a failure to connect the problem to the overall strategy. In the absence of a clear strategy that provides context, many solutions appear to make sense.

6. **Over-dependence:** Some decisions are never made because one person is waiting for another, who in turn is waiting for someone else's decision or input.

7. **Failure to communicate the what, where, when, and how associated with their decisions**: Some good decisions become bad decisions because people don't understand – or even know about – them. Communicating a decision, its rationale and implications is critical to its successful implementation.

(Zenger & Folkman, 2014)

Most companies I've worked for have written thick tomes around their culture and values. It results in a manifesto that becomes a massive paperweight because no one ever reads it. I was starting to get a sense that this company was a bit different. It had just six core values: Open, Authentic, Inclusive, Collaborative, Creative and Tenacious. I think what made it different was that instead of paying lip service to them, I was seeing people at all levels walk the talk.

Shortly after I started, they sent me to two separate training programs. The first was a basic three-day new hire program. You focused on getting to know the culture and values of the company. They wanted you to know how the leadership team behaved and what motivated them. They put us through a leadership program called Reality Based Leadership. It was derived from the work of Cy Wakeman in her best-selling book based upon those same principles. The premise was simple yet I had not heard of it before: take the drama out of the workplace (Wakeman, 2010). Workers should focus on doing good work and take the drama out of their daily interactions with others. It was Forrest Gump-simple, but it totally works.

We also did a personality analysis. Understanding which way you 'tilted' would help you in terms of your interactions with peers and extended team members (Tilt, 2019). It gave you a good perspective on how and why people behave the way they do, and provided a path towards empathy. Sitting in class that day, I observed a moment of silence. I was floored. All the companies I had worked at before had focused on technical skills. They spent weeks showing you how to do your job. They provided training on the nuts and bolts of using their tools, but they left the people aspect to its own devices. At this company, it was completely reversed.

Two weeks later they sent me on a week-long training program in San Francisco. In the heart of the Valley I met 50 of my teammates from around the world. Their functions were wide-ranging: marketing, sales, operations, and even executive team members. No one was exempt. Every new hire had to complete this training. Every day they hosted executives and subject matter experts to immerse us into the culture of the company and show us what exceptional looked like.

At the end of the week, I was completely energized. To be clear, I'm not sure I had enough information and training to do the nuts and bolts of my job. But what I received, I think, was more important. They knew they were in hypergrowth mode and didn't have all of the people or processes figured out yet. What they wanted to do was hire really smart people and let them be part of the solution. My 50 cohorts and I would be part of the vanguard that would help this company achieve its goal of becoming a $1B software company. That's why providing guiding principles was more important than the details on how to do your job.

That night, as we were having dinner, my wife asked me my impressions about the new company. I reflected on the last two months. It reminded me of my first two months in university. You are in a completely new environment. There are really no rules or guidelines. The professors didn't care if you attended class. The method and medium in which you absorbed the required material is up to you. Whether you pass or fail the exams is totally dependent on your efforts and desires. I gave her the most honest answer I could:

"I have no idea if this is going to work out. I like what I see so far. I love the culture. The people that I'm working with seem to be really nice. I'm going to focus on the only thing that I can control – I'm going to work my butt off and see where it takes me."

Lessons Learned:

➢ There is no right or wrong when it comes to choosing the size of company you want to work for. There are pros and cons to both. Think about what you personally prefer and what you need to be successful. If you like structure, hierarchy, and processes, then you might prefer a big company. If you like a manic pace and not many rules, you might prefer a smaller company. Smaller companies can't provide all of the benefits that a bigger company can afford. The key question you have to ask yourself is: Do you really need those benefits at this point in your life?

➢ If you have been in the corporate world for 10-20 years and have worked for mainly big companies, take a risk and work for a small company. It will give you a completely different perspective. It will give you new muscles and skills that you never knew existed.

➢ Companies have different stages of life. Figure out which stage you like working in the most and target companies in that stage. When the company passes that stage, it might be worth it to find the next wave to ride.

CHAPTER 12

Talking To The Wrong People

5:00 am: Wake up. Start researching. Drink lots of coffee.

7:00 am: Take a break. Have breakfast. Shower.

7:30 am: Start cold calling. Leave voice mails.

8:30 am: Start emailing prospects.

10:30 am: Take a break.

10:45 am: Continue research on prospects.

12:30 pm: Break for lunch.

1:00 pm: Start cycle all over again.

6:10 pm: Call my therapist.

"Jonesy, hey bud, what are you doing? OMG. I can't take this anymore. My back hurts, my butt hurts and I can't look at my computer anymore. I'm completely bug-eyed."

This was my daily call to my peer and friend, Jonesy. Jonesy and I were on the same team and had the same mission. We were going to crack some of the largest global customers in the world. Nearly all of the accounts we were calling on were prospects – that meant that they were not existing customers and it was our job to change that. Jonesy and I were going through the exact same thing, so it was therapeutic for us to talk to each other.

I'm not sure if Jonesy wanted to talk to me every day, but it was way cheaper than seeing a therapist. Every day at some point, I would call Jonesy and tell him my "woe is me" story. As the days turned into weeks and the weeks turned into months, I started noticing a disturbing pattern. When I started my prospecting campaign, I would call Jonesy at 6 pm or later to get my dose of positive affirmation. After two months, I caught myself calling Jonesy at 10 am one morning. You know it's not good when that happens.

Two and a half months later.

"Trong, this looks interesting. Let's set up a meeting to discuss." Did Christmas just come in March? I had to read the email multiple times. I couldn't believe it. One of my prospecting messages had resonated with the Global CFO of the pharmaceutical customer I was calling on. And he wanted to meet! Forget my wife and family – the first person I wanted to tell the good news to was Jonesy.

Within an hour of the Global CFO's email, his executive assistant reached out and we set up time to meet three weeks out. I was on cloud ten. My jubilation lasted only a week. I received an email from the Global CFO's SVP for Finance.

"Trong, we already reviewed your solution a year ago. We found it didn't meet our needs. There is no need for you to meet with the Global CFO." Carrie didn't waste words. I didn't know her, but I didn't like her already. In three sentences, she crushed my dreams.

Prospecting is like exercising: it is not for the faint of heart or the uninitiated. You have to do it every day to build up the stamina and muscle memory to be good at it. The constant rejection can take its toll on anybody. However, if you don't do it, you will never become good at it. There is evidence for what will resonate with prospects and how to improve your odds of getting responses from executives. Below are some interesting statistics from the Rain Group (Rain Group, 2019):

1. Eight in 10 prospects prefer talking to reps over email.

2. Half of buyers like speaking over the phone, compared to 70% of reps. That percentage increases the higher up the ladder you go. If you are working with a VP or an executive, it is best to give them a call.

3. Prospects are open to communicating with sellers at industry events (34%), via LinkedIn (21%), text (21%), voicemail (21%) and on social media (18%).

4. The biggest reasons a prospect will talk to a vendor are: they have a need for your product (75%), they have a budget (64%), they think you have value to offer (63%).

5. For C-Level executives, 75% of them will take a call because of an ROI business case (vs. 64% for directors and 59% for managers).

6. It takes 18 dials to connect with a single buyer.

7. Call back rates are less than 1%.

8. Less than 24% of sales emails are ever opened.

9. Leads responded to within 5 minutes are 100X more likely to be qualified.

10. Only 27% of web-generated leads get contacted at all.

11. Your sales team has a 56% greater chance to attain quota if you engage buyers before they contact a seller.

12. The first viable vendor to reach a decision maker and set the buying vision has an average close ratio of 74%.

13. 50% of buyers choose the vendor that responds first.

14. 73% of executives prefer to work with sales professionals referred by someone they know.

15. 84% of B2B decision makers start the buying process with a referral. (Schultz, 2018)

Three days later.

Instead of moping and wallowing in my own business development misery, I decided to resume my cold calling and prospecting. I continued my daily ritual of calling Jonesy just to tell him how much my lower back and butt hurt from sitting on it all day. Another day, another therapy session.

One week later. 6:06 am. An email from Carrie.

"Trong, I told you we are not interested in your product. Our technical teams met with your technical teams and we determined your products didn't meet our needs. Please stop emailing. And stop calling my boss. We are not interested."

I stared at the email for a good ten minutes. It was definitely harsh, but not unwarranted. I had done everything that Carrie had accused me of. I had continued to call and email her boss after she told me they had already reviewed our solution and found it lacking. I was going to need sweat therapy to get over this one. I decided to go to the gym and run on the treadmill for a good hour.

8:01 am I decided to call her and address this issue head on.

"Hi Carrie, this is Trong. I got your email this morning. I'm confused. Deep down, I know that our solution has everything that your company is looking for. I know we could help solve your business issues and make your company much more agile than it is today." I was transparent and honest. I wasn't going to play any games.

"Trong, I had already said this in my email. My technical teams talked to your technical teams and we determined that it wasn't going to meet our needs. There's not much to say beyond that." Carrie was being direct and transparent. I didn't get a sense she was playing any games either. I was starting to like her.

The gears were spinning in my head. I paused and thought about what she had just said.

"Carrie, I think I know what the problem is. If my technical teams are talking to your technical teams, then we are talking to the wrong people. Our solution is designed from the ground up to be a solution that is owned and managed by the business user. It was never designed to be a solution for IT. Let's do a reset. Let me walk you through what our solution actually does and what it is meant for."

With that, I went into a 25-minute discourse on our product and solution. I was regurgitating everything that I had learned from our Sales Enablement team. Carrie listened intently and absorbed everything that I said. She was smart. She was tracking with me.

"Carrie, that is a lot of information for you to digest. Why don't you think about what we've discussed and we can reconnect in two weeks? I know there are things we can do with your teams to remove some of the manual processes from your organization."

"Trong, I think that's a great idea. Let's find some time in two weeks to see where we go from here."

Two days later. 4:30 am Another email from Carrie.

"Trong, I couldn't sleep and was thinking about what we had talked about. We have this problem ..." With that, Carrie went on to describe their current predicament and asked if our solution would be able to solve their business issue. Is bacon a vegetable? Is Santa Claus real? Is Hawkeye an Avenger? Absolutely!

"Carrie, our solution can do everything that you are looking for. This is exactly what it was designed for. I'd like for my systems consultant to come in and give your business team a demo of our product. Since the solution is designed to be owned by them, they should see it in that context. They would be the best people to evaluate if this product is what you are looking for."

I couldn't wait to call my therapist – Jonesy.

Lessons Learned:

➤ Prospecting and business development is hard. You won't be good at it unless you practice it regularly. It can be very emotionally draining. There are things you can do to recharge the batteries along the way: exercising, taking multiple breaks, and calling a good friend/peer/colleague just to let off some steam.

➤ If you find yourself calling your "therapist" earlier and earlier every day, it means you are mentally and emotionally burning out. It is probably time to switch things up and do other things besides pure prospecting.

➤ When you are trying to break into a prospective account, target high. If you are trying to affect change, only the top leaders can do that. Middle management may be resistant to change. Call executives early in the morning, at lunch time, or after business hours – you'll have a higher probability of getting them live than during regular business hours.

➤ Customized emails and targeted prospecting will yield better results than generic mail blasts. Don't think that your emails just get lost in the ether. If they are done right and there is substance to them, they are being read by the executives that you are sending them to. They might not respond to you, but they are being read.

➢ When dealing with executives, be direct and transparent. They will appreciate your honesty and conciseness, as they do not have a lot of time to waste.

➢ Be persistent. If you know something to be inherently true, then it probably is. If your message doesn't seem to be resonating, it just means it hasn't found its way to the right people yet.

Trong Nguyen

CHAPTER 13

Mr. Pink And Reservoir Dog

∽∾∽

Breaking up is hard to do. I got mesmerized by the shiny new object. He looked slick. He talked slick. He said all the right things. He made me feel good about myself. He pumped me up. I felt complete when I was with him. In my mind, I flashed forward and thought we could have a long future. We were going to build a house with a white picket fence and close so many deals together.

Then one day, it all fell apart. Why didn't I see it coming? I wanted it to work so bad that it hurt. When the going got tough and I needed his help, he was never there. He made promises of teamwork, comradery and esprit de corps. I cherished his every word and would do anything he wanted me to do. I would charge any hill, take down any sale, and recite his orders like commandments. When I found out that they were just hollow words, my heart broke. When you find out that your manager is not who he says he is, you have to pause and reassess the situation. Do you really want to be in this relationship? For me, there was a line you couldn't come back from.

In the valley of despair, I kept on humming and singing the George Michael anthem in my head. Not the Anna Kendrick version that kids see on Pitch Perfect today, but the original one.

I just hope you understand
Sometimes the clothes do not make the man
All we have to see
Is that I don't belong to you
And you don't belong to me, yeah yeah
Freedom! Freedom!

It took me a few months, but I decided to find a new leader to follow. When you go through an emotional situation, it can lend itself to extremes. If you lost a lot of weight, you might start binging on food. If you dated a vegetarian, you might start eating lots of bacon. My situation was no different. After my breakup with Mr. Slick, I decided I didn't want to be with slick anymore. I wanted to be with someone dull and boring. That's when I decided to hook up with Mr. Pink. Seriously! He looked exactly like Mr. Pink in Quentin Tarantino's film *Reservoir Dogs*.

Mr. Pink was paint-dry dull. He didn't dress to the nines and he didn't make grandiose promises. He wasn't a rah-rah-rah leader in the classical sense. He was so chill, sometimes I wasn't sure if he was alive. He reminded me of someone's grandfather. He seemed balanced. A perfect fit for my manic self. Mr. Pink had been around the block a few times and had the scars to prove it. He told stories of how Marc Benioff shaped Salesforce while he was right there in the middle of the action. He wasn't bragging. He was just stating facts and telling you how it really was back in the day. He hired me in the hopes of building a new financial services vertical. If it worked out, I could see us building a sales house together – with a white picket fence of course!

Mr. Pink and I didn't have a lot in common, but there was something that I liked about him. I just wasn't sure what it was. Then one day it hit me. **Character.** He had character! Mr. Pink talked about his wife and grown up kids, and every time he did, he gushed. You can tell that they meant the world to him and they were his first priority – even more important than work. Most executives say that, but you don't see it. They preach that family is everything to them, but then drop their families like a hot potato when it is inconvenient for them. Mr. Pink was different.

Sometimes I like to send up a trial balloon just to see how someone would react. Spies in the cold war would do this all the time. They would spread misinformation and see where it would lead. In an indirect way, these trial balloons can tell you everything you want to know.

"Mr. Pink, be honest, do you want that deal to close this quarter or next quarter? If it closes next quarter, you get the credit, right? If it closes this quarter, the old VP gets the credit, right? I'm thinking, it is way better for our new team if it closes next quarter."

There it goes. I could hear Nena singing in my head: "99 red balloons, floating in the summer sky…"

"Trong, as much as I want the credit for the deal, I don't want to be that guy. I didn't do any work for it. I hope it

124

closes this quarter so the old VP can legitimately get credit for it."

There it is – **character**!

Three weeks later we were on a plane to San Francisco for our global sales kickoff. I got to meet some talented new teammates. On the last day, our SVP of Sales talked to us about the importance of getting a fast start, working hard and having an open mind. He asked all of us to read Carol Dweck's *Mindset: The New Psychology of Success*.

The book is riveting. I finished it in one day. As I turned the pages, it was a mirror of my life to date. I had made every mistake they outlined in the book and then some (Dweck, 2006). Here are some of my notes:

1. In 264 pages, I saw all my failings right in front of me. It was insight I wish I had in my twenties. I had too many lessons learned the hard way and scars earned through a lot of pain. Hindsight is 20/20, but I wonder if I would have behaved differently in certain situations if I had this knowledge ahead of time. Or is it more of a function of maturity and getting older and wiser?

2. I really liked the chapter on teaching your kids about effort vs. results. Train them to try their best and reward that rather than the end results. This gives them the tools to cope with adversity later on. As a father of three young kids, I found this advice to be invaluable.

3. I liked the chapter on owning your destiny. Too often we find others to blame for our failures. If we look inward and force ourselves to own our destiny, then we will always be the best version of ourselves. The story about John McEnroe is really telling. I knew he was a fierce competitor and had a bad temper, but I didn't know the back story.

4. As a business buff, I spent my formative years learning about Jack Welch (I met him in business school) and Chainsaw Al. I remember when Chainsaw Al became CEO of Sunbeam. I'm not sure I agree with the premise that Chainsaw Al had a fixed mindset. He was more of a fraud than anything else and that's what caused his demise at Sunbeam and the other companies he ran. He was too loose with his accounting practices and made Enron look squeaky clean. Being a fraud and

lacking a moral compass is a bit different than having a fixed mindset.

5. I ran cross country and track in high school. My coaches were always pushing me to just beat my personal record. That's what they were interested in. I never really understood that concept until now. The growth mindset is the same mentality: If you continually do incrementally better and try to beat your personal best, then you will ultimately be the best version of you that you can be.

When you have been in sales for as long as I have, you get set in your ways. You rinse and repeat your formulas over and over again because these methodologies have made you successful over time. To a certain extent, you develop a fixed mindset – that's why it is hard to teach an old dog new tricks. And here I was in my prime, an old dog.

I decided to change. Like everything I have ever done in my life, I dove in head first. A lifelong Microsoft devotee, I ditched my PC for the Mac. I gave up the mouse for a trackpad. I gave up bacon for a healthier lifestyle. OK that's a lie! I was just making sure you were paying attention. I decided to embrace the growth mindset and see where it took me.

After a few months, Mr. Pink and I were waist deep in a sales pursuit. At this point I had closed so many big deals that I was cocky and full of myself. I wrote the book on enterprise sales and I wasn't going to take feedback from anybody. My proven methodology worked and had made me a lot of money. I was on a status call with Mr. Pink when he made a suggestion out of left field. It was a competitive pursuit and he suggested we add some humor to the meeting by translating our point of view document into Italian. Our chief competitor's support organization was all based in Italy and their support documentation was all written in Italian. I thought to myself, Mr. Pink has totally lost it. He should retire. I knew he was old, but dementia doesn't kick in that early, does it?

I paused. I took a deep breath and then decided to try Mr. Pink's suggestion. We went into our meeting with the customer. We opened it up with the formal review and how we stacked up against our competitor. Then I closed by sharing the documentation translated in Italian. The customer laughed so hard I thought he was going to cry. In that instant, we had shown our prospective customer that we had the depth and content, but that we also had style and personality. That was the turning point in our sales pursuit.

After that I talked to Mr. Pink every day. He spent most of the time listening, but when he spoke, it was gold. Even when his advice sounded wacky and was something I wouldn't normally do, I decided to do it anyway. He was always right.

On the last day of the quarter, we were still chasing our big deal. This deal meant a lot to our team, as well as the region. It was also significant because we would be signing a premier insurance company as a new customer. Mr. Pink came to the customer meeting with me and played the executive sponsor role. He was right there in the trenches with me. We decided to park ourselves near the customer office so that we could put out any fires in real time. When you are down to the final strokes in getting a deal done, everything is amplified a thousand-fold. The difference between bringing a deal in the current quarter as promised vs. two days later is the difference between a Super Bowl champion and everyone else.

What I found out about Mr. Pink was that he had a unique ability to think very strategically, but he was just as adept at diving into the details and making sure we didn't die by a thousand cuts. He had every administrative action covered. Near the close of business on the last day of the quarter, our purchase order came in. We jumped for joy and hugged each other silly. We were both ecstatic and relieved at the same time. Mr. Pink had taught this old reservoir dog some new tricks.

Lessons Learned:

➤ Embrace a growth mindset. It will allow you to become a better version of yourself faster. True winners in any field are life-long learners. They are always trying to learn different ways to improve their craft.

➤ Business relationships are like personal relationships: You go in with the best intentions and try and make it work, but if it doesn't work out, it's better to acknowledge it and move on. Two wrongs won't make a right.

➤ Character and integrity are worth their weight in gold. If you are around people without integrity, run far away as fast as you can. One test that you can easily apply if you are unsure is the parent test. Would you be ok to introduce this person to your parents and vouch for their character and personality. If you aren't, then don't hang around them or work for them.

➤ When you can, inject some of your own personality and style into your customer interactions. That is what will differentiate a top seller from an average seller. Instill some humanity in your business dealings.

Trong Nguyen

CHAPTER 14

How I Barely Survived A Pandemic

Dostoyevsky, Van Gogh, Beethoven, George Orwell, Tennessee Williams, and Kurt Cobain (Nirvana): What do they all have in common? They are creative geniuses who were the voice of their generation. They defined their particular crafts for their time. Their output was so profound that decades, and even centuries later, we are listening to, looking at, and savoring their work as if it were a well-aged wine. What is striking about these artists is that they all had tortured souls. They were running away from something. It could have been their upbringing and background, their relationships, their physical deformities, or even their station in life. Whatever the case, they were in deep pain. They dove headfirst into their craft as a form of escapism and turned their raw pain into artistic gold. The idea that "great art comes from great pain" (Zara, 2012) has been a long-held belief in society.

Unfortunately, this belief turns out to be incorrect. There have been numerous studies that have shown that there is no correlation between extreme pain and suffering and great art. Over time, we as a society have romanticized and glamorized these notions to mythic levels.

Our modern idea of the brilliant, tortured genius likely stems from a glamorization of mental illness that took hold during the Romantic Era (Bianca, 2018). In reality, there is exponentially more memorable music, books, and art produced by talented and happy artists.

When you talk to most veteran sales managers and ask them what they are looking for in a sales rep, they often answer with the cliché that they want to hire a sales rep who has an expensive lifestyle and is in serious debt. Their belief is that those people have the most motivation and will work their butt off to make their quota because they have massive financial incentive to do so. This is similar to the "tortured artist" belief, except now it's the tortured sales rep. The more turmoil, personal debt, and financial need they have, the more incentive they have to work harder and find every which way to achieve their quota. Making their quota means everyone wins - this includes the sales rep, management team, and corporation. Could sales managers be onto something? Could the folks that did the studies at HBR be wrong (Mayer & Greenberg, 2006)?

March 13, 2020

We weren't sure if we were being reckless and irresponsible. We weren't breaking any laws, but it sure felt like it. We were at the airport waiting for our plane to board to Florida. This was our planned family winter getaway. We were done with winter. We wanted to bask in the Florida sun to get away from our gray, dreary existence.

After we checked in, we noticed that the airport was eerily quiet. I've travelled a lot over the last decade and this is the first time in my life that I had ever seen it this quiet. Normally the airport was bustling with business travelers and vacationers. Happy people trying to get to their destinations and sad people running away from their problems. And now, there was just a low murmur. A few stragglers here and there. Almost crickets.

News was starting to break out that the new coronavirus was spreading at a rapid pace. The government at different levels had started to encourage people to stay home if it was at all possible. There was a new concept that was being bantered around called "shelter in place." That meant that everyone should stay at home and avoid contact with anyone else for fear that we would unwittingly spread the disease.

The first day in Florida was fun. We went to the beach and soaked in the sun and Atlantic breeze. Unfortunately, the next day the county mandated a shutdown of the beach and all public gatherings. That meant that even the pool and common area in the condo that we rented was going to be shut down as well. We went home and watched the news to gauge just how bad the situation was. States all over the country were starting to adopt "shelter in place" policies. There was even talk that they would start shutting down the borders between the USA, Canada, and Mexico.

March 16, 2020

Pandemonium started breaking out everywhere. Hand sanitizer, toilet paper, flour, yeast, and chicken thighs were completely gone. People were hoarding them and thinking that we would be locked down for the next couple of weeks. With that as a backdrop, we decided to cut our losses and go home. We had just spent a lot of money, but vacations are not really fun when you are in a pandemic.

Week 1-2

I loved being locked down! I slept in until 7 am every morning and then rolled myself into the home office. I was at work by 7:05 am. I had dinner with the family every night, and we topped it off by playing fun family games and doing puzzles. In two weeks, I had put on 5 lbs. This was the most sleep and the most time I'd spent with my family in a decade. I had no idea what people were complaining about. Social distancing wasn't all that bad. I was going to enjoy this time with my family for as long as I could.

Week 3-4

We were all adjusting to the new normal. Instead of in-person customer meetings and brainstorming sessions where we were whiteboarding and having multiple parallel conversations, we resorted to WebEx and Zoom calls. A few drinks after work with colleagues? Those were gone. We replaced them with Virtual Happy Hours. Visiting parents and extended family? That was gone as well.

Actually, maybe that one wasn't so bad after all! Everything became a WebEx or a Zoom call. You don't realize how much you socialize and interact with people on an informal basis until it suddenly stops. Being locked down wasn't fun anymore. I had put on 10 lbs. and was eating out of sheer boredom. Customers were trying to deal with the new normal as well. Projects and deals were being pushed out months and even quarters now. This was definitely not good.

Week 5+

I cracked. I'm not sure how it crept up on me, but I knew it when I saw it. I snapped at an innocent family member for something so trivial it was ridiculous. Whether I wanted to admit it or not, I was missing people. I missed the social interactions, the comradery and the touch. And I missed being with my customers. Being in sales and not being able to physically meet and be with your customers is complete TORTURE. This is the pain and turmoil that those artists must have felt.

How do prisoners not go bonkers being locked up? How did Nelson Mandela survive 27 years of isolation? It's insightful what you find out about yourself when you are isolated and not able to physically be with other people. I discovered that I'm not as strong as I thought I was. The realization that I would make a horrible prisoner and that I was cracking under the strain of a pandemic was a blow to my brittle soul.

In my darkest hour, when all I could think about was sucking my thumb in a fetal position, a speck of light appeared in the distance. I started to smile. I couldn't paint, compose music, or write a Pulitzer prize winning book but I could become a better salesman. I promised myself to turn this pain and grief into something good.

If we made it through this pandemic, I was never going to take things for granted again. I vowed to cherish every customer meeting and interaction. I would jump at any opportunity to solve a tough customer problem. I would treasure every impromptu meeting at the water cooler and social interaction in the client cafeteria. I was going to use this pandemic to make me a better salesman.

I started to smile because I realized that I was one of the luckiest guys in the world. Selling game-changing software to the largest companies on the planet is a privilege and not a right. I was going to enjoy every moment of it. Baby steps. One day at a time. You can make it through this...

Lessons Learned:

➤ Selling to customers virtually is really hard. The customer interaction via video calls is not the same as in-person meetings. It is hard to run brainstorming sessions via video calls because in that medium only one person can speak at a time. You can't have multiple parallel conversations.

➤ We don't realize what social creatures we really are until we can no longer socialize. Treasure every minute that you have with your customers, co-workers, family, friends, and strangers because one day, you may find the lack of interaction hard to take.

➤ If you had a chance to do a reset, what would you do differently? How would you behave? Who would you surround yourself with?

➤ Everyone has a different story, different support structures, and different home lives. Respect, empathize, and help where you can.

CHAPTER 15

Unicorns Do Exist!

∞

I blushed every time I came into the office. I had no idea why I was so giddy. I had just joined this small software company and immediately got a sense that it was very different. As I looked around the room, I saw a female Asian sales rep, another male Asian sales rep, and then Kim, the Regional Vice President. I have been in sales for over 20 years and I had never seen another Asian in a sales capacity at a tech company before. Every Asian I'd met in a professional capacity was in a stereotypical role: the math genius in finance or the brilliant technical wizard writing software code that made everything work. The first time I met an Asian in a leadership position was at IBM. Harry was the head of IBM Research. The second time I met an Asian leader was a decade later. He was the head of accounting for a Canadian bank. Kim was the third.

I smiled every time I saw Kim. If I was straight out of sales school, he would, without a doubt, be my hero and role model. He would be everything I was looking for. Kim was tall, impeccably dressed in his skinny runway suits, and technically brilliant. If that wasn't enough, Kim could sell a master class in value investing to Warren Buffett. His customers loved him, and his team loved him even more. An Asian VP in tech sales. I couldn't believe it. I finally found my unicorn and his name was Kim. Why weren't there more Asian sales VPs? Did we not have the right skills? Did we not have enough talent? Maybe we just didn't want to be VPs?

One year later.

The COVID-19 lockdown had lasted for almost three months now. If you had told me when this all started that we would be locked down for this long, I would have called you silly. Emotions were starting to fray for everyone across the country. The tension was thick as molasses.

"Dad, did you see that video on Amy Cooper? It's gone viral."

"Who's Amy Cooper?" I had no idea. My kids and all of their friends were watching this video of a white woman who falsified her situation when she called the police on a Black man while he was bird watching. As I watched the video, the hair on the back of my neck started standing up.

This video came on the heels of the footage of George Floyd breathing his last breath in police custody in Minneapolis. Eight minutes and 46 seconds. The police arrested him, put him on the ground in a choke hold, and held a knee on his neck for eight minutes and 46 seconds. My heart shattered. The nation erupted.

What followed was massive civil unrest. Protests started in Minneapolis and other major cities across the U.S. People organized around the world. There was so much pent up anger, rage, and frustration. Black Lives Matter came back into the spotlight, and we were finally talking about the need for significant change in systemic racism, white privilege, and police reform: issues so divisive that friends and neighbors across the country were in constant heated debate.

As we were bombarded by media coverage, I stopped to reflect on why people were so enraged. I thought about how systemic racism affects ALL economic classes of minorities, but has a disproportionate effect on those experiencing poverty. I pondered my own situation: I was a minority who also came from abject poverty, but my personal experience led me to naively believe that if you worked hard, anyone could live the American dream. I started to ask myself, what was different? Why was I able to break through the economic and social glass ceilings when others couldn't? Weren't the opportunities that I had available to others? As I started to explore the intricate relationships between economics, power, and social status, I had an epiphany.

Financial textbooks define net worth as:

Net Worth = Assets - Liabilities

A 2016 study by the Survey of Consumer Finances finds some eye-opening discrepancies underlying that simple calculation. They found that the racial gap in net worth between white households and Black households was a factor of 10.

Median Net Worth by Race

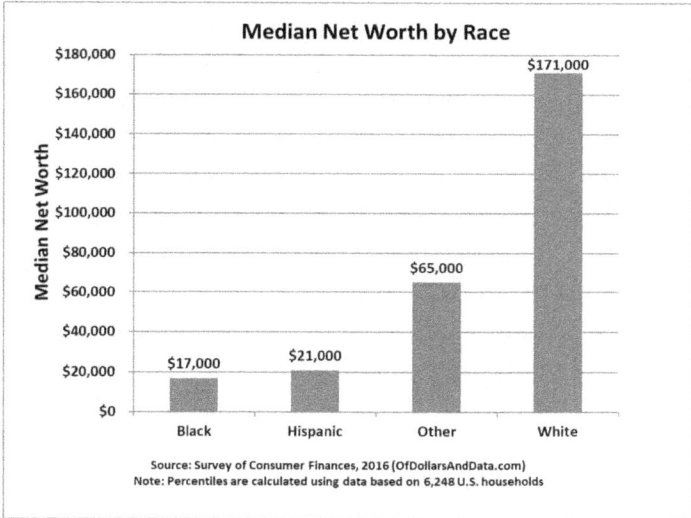

Source: Survey of Consumer Finances, 2016 (OfDollarsAndData.com)
Note: Percentiles are calculated using data based on 6,248 U.S. households

(Maggiulli, 2020)

A typical white household has a net worth of $171,000 and a typical Black household has a net worth of $17,000. You might be thinking the same thing I did: education is the ultimate equalizer. Education helped me break through the economic barrier. I thought that if other minorities became better educated overall, they could do the same thing I did.

If you peel back the onion a bit more, you find that currently a college degree barely gets a Black household past where a white household is with no high school education (Maggiulli, 2020).

	Median Net Worth - 2016	
	No Bachelor's Degree	Bachelor's Degree
White	$98,100	$397,100
Black	$11,600	$68,200
Hispanic	$17,500	$77,900
Other	$34,300	$210,200

Source: Federal Reserve Board, Survey of Consumer Finances

What made my situation different? It was the power of compounding! I was a refugee who got sponsored by the Catholic church. They sent me to an all boy's Catholic school. I was fortunate enough to get a top-notch education. I went to a highly respected university for undergrad. I got an enviable job straight out of school. I received life-changing mentorship and guidance from my bosses. And this continued on for another decade with the different companies that I worked for. One advantage after another after another. That was the power of compounding. It wasn't just one thing. It was a series of advantages compounded over 30 years that allowed me to break through the social and economic barriers.

I learned that for centuries, Black people have been mistreated, ignored and neglected. When you have such a disparity between the haves and have nots, inequality becomes ingrained. We can no longer have some with advantages and privileges while others do not get access to the same resources. We can no longer just watch this happen, reap the benefits of our privilege, and say nothing. We must now think and act.

As I reflect on everything that has happened, I started to think about my hero, Kim. It's easy to assume that he started with numerous advantages similar to mine, and they kept on compounding. What I hope is that over the next decade, we will see more Asian, Black and other people of color sales leaders because systemic racism has been altered to allow a more even distribution of those early advantages that can be compounded over time by society and caring individuals. I must invest my own energy to this cause. I can no longer stay silent. I want to do my part, and it starts with understanding my own privilege and my own abilities to contribute as a positive force in the educational journey of racial minorities. Contrary to what many believe, unicorns do exist, and they come in every color.

Lessons Learned:

➤ To truly help the disadvantaged, you need to help them over a sustained period of time. There is no silver bullet or short cuts to fixing inequality.

➤ If you want to see more unicorns, invest actively in change now and then be patient. The power of compounding will bring returns in time.

➤ We have an obligation to help those less fortunate than we are. Society is a better place when everyone wins.

➤ It melted my heart when my kids brought up the issues around the Amy Cooper and George Floyd incidents and why they thought it was so wrong. They are our future unicorns. Their perspectives and insights will change the world.

➤ The issues around diversity and inclusion are bigger and broader than the simple story I have just told. If you zoom out a bit, why do you rarely see a person of color as a VP of sales? You may see them as first line sales managers, but even that is exceptional these days. Why is that? Why are most women execs in purchasing or marketing? Why are most Black execs in HR or chiefs of diversity? Why are most Indian execs in technology or IT? Why has that not changed in 30-40 years?

CHAPTER 16

Looking For Salvation

SAITO: If you can steal an idea from someone's mind, why can't you plant one there instead?

ARTHUR: Okay, here's planting an idea: I say to you, "Don't think about elephants."

(Saito nods) What are you thinking about?

SAITO: Elephants.

ARTHUR: Right. But it's not your idea because you know I gave it to you.

SAITO: You could plant it subconsciously.

ARTHUR: The subject's mind can always trace the genesis of the idea. True inspiration is impossible to fake. An idea is like a virus. Resilient. Highly contagious. And even the smallest seed of an idea can grow.

Christopher Nolan will arguably go down in history as one of the very best filmmakers of his era. I've been a huge fan of his since his early success with mind-bending movies such as *Memento* and *The Dark Knight* trilogy. One of my all-time favorites is a movie that he wrote, directed, and produced called *Inception*. This is a complicated heist movie with the premise that you can implant thoughts and ideas into people's minds through a process called "inception."

Early 2000s.

"Trong, why don't you find a hobby. Something to broaden you out a bit. You are always working. That can't be healthy for you."

Michelle and I had been married a couple of years now. I know she was speaking English because that was her native tongue, but she could have been speaking alien to me. I had no idea what she was talking about. Hobby? Stop working so much? The next thing you know, she's going to start asking me to volunteer. Doesn't she know volunteering doesn't pay?

Fifteen years later.

"Trong, why don't you find a hobby. Something to broaden you out a bit. You are always working. That can't be healthy for you."

Inception. There it is. It only took fifteen years for that spark of an idea to blossom, but I guess it is better late than never.

I started to noodle on what Michelle was saying. I'm not sure if I was just going through a mid-life crisis or if I finally had some semblance of maturity.

"What do you think I should do?" I was starting to open up to the idea of doing something else besides work.

"Well, you are good at computers and writing. Why don't you volunteer somewhere where you can help kids with computer skills or their resumes? Do that maybe once a week. I don't think you have more time than that to spare."

I liked where this conversation was going. Michelle did have a point. We had been good savers all our lives. Thankfully we were not living paycheck to paycheck anymore.

Professionally, I was at the top of my game. I still loved sales and was having a blast. Not everyone can say that about their job. Maybe the time was right to pay it forward and give back. I couldn't do what Michelle was suggesting I do though. I don't think I could just sit back and volunteer a day or two. It goes against the grain of who I am or how I would approach things. If I was going to do some charity work or volunteer, I had to be all in. I couldn't just do something for a few hours a week and leave it at that.

In researching non-profits and charities, I came across some interesting statistics:

- 31% of donors worldwide give to NGOs, NPOs and charities located outside of their country of residence.

- 41% give in response to natural disasters.

- Giving to education charities was up 6.2% to $58.9 billion (14% of all donations).

- In 2017, the largest source of charitable giving came from individuals at $281.86 billion, or 72% of total giving, followed by foundations ($58.28 billion; 15%), bequests ($30.36 billion; 8%), and corporations ($18.55 billion; 5%).

- 30% of annual giving occurs in December.

- 77% of the public believe everyone can make a difference by supporting causes.

- 64% of donations are made by women.

- 69% of the population gives.

- 55% of people who engage with nonprofits on social media end up taking some sort of action.

    (National Philanthropic Trust, 2019)

But what could I do that would help save my soul as well as help kids in need? I started thinking about my childhood. I was dirt poor and it was the generosity of a handful of people that completely changed the trajectory of my life. These saints and good Samaritans provided funding so that my sisters and I could go to the best Catholic schools and get a top-notch education. That education enabled me to go to the best universities. And from there, I was able to get the best jobs. It was a snowball effect that started with access to a quality education. This was the genesis and idea behind the non-profit I decided to create.

"You dummy! You totally misinterpreted what I said! I told you to volunteer. Start doing some good in the world. Take a small step in the right direction. I didn't tell you to start your own non-profit!"

"It's all your fault! I told you I only had time for two things. Let me just focus on work and family. I knew I didn't have time for anything else!"

I was trying my best to spin this and shift the blame back on Michelle. I had no idea how I got myself into this mess. I was spending every minute on the weekends now for the last month trying to set up my non-profit as well as come up with the creative and content for our website. I was burning the candle at both ends. I'd work 70 hours a week at my day job. At nights and on weekends, I did all the prep necessary to spin up my non-profit so that I could help kids in need. I had no idea giving could be so hard!

After the dust had settled and I started to think rationally again, I knew it was all my fault. Michelle did tell me to just volunteer and help out a few hours a week. I took that encouragement and did what I always do with an idea: I took it to an extreme. Everything was always so binary with me. I was either all in, or not in at all. In this case, I decided I was all in.

There was no turning back now. I had already committed to the idea of starting a non-profit. I had roped in one of my best friends, Kathy, to be the treasurer and CFO. I had asked two other good friends, Stephanie and Adam, to help build the website. I was in daily contact with the IRS, providing supplemental information so we could register as a $501(c)(3)$ charity in the USA.

Once you are knee-deep in something, the only way out is to move forward as fast as you can. I wasn't looking back. Kids all over the world needed our help and we were going to give it to them. When faced with a daunting task, I become hyper-focused. I break things down into manageable chunks and start executing those tasks to closure. This would be no different. Break it down. One piece at a time. You can do this.

Saturday morning. 5:14 am.

While the whole family was sleeping, I was in the kitchen, sipping my black coffee. I was cleaning my potential donor list. Over the last twenty years, I had amassed over 11,000 contacts. I combed through this list and came up with about 300 people who would be my initial cold calls. These were friends and close business associates that I knew I could call on to help me make a difference. I only had a few more things to do before I launched the campaign. All the hard work over the last few months was starting to come together. I felt nervous. I felt excited. I didn't know what to expect. Was I going to change the lives of some deserving kids? Or through this experience of giving, were they going to change my life and make me a better person? I'm thinking it is probably the latter.

Lessons Learned:

➤ The biblical idea of *Give and You Shall Receive* is totally true. Paying it forward is not only good for the recipient, but it is can make you feel good as well.

➤ When someone gives you good advice, be smart enough to listen to it. Don't let it linger for fifteen years unnecessarily. The next time someone tells you something you disagree with, try this exercise: Take the message at face value and extract all the emotion out of it. Does it make sense? If it does, then the odds are high that it is good advice and you should listen to it.

➤ It is alleged that balance is a good thing. You don't always have to take things to extremes. I can't authoritatively comment on this as I have not been a good practitioner of balance.

Trong Nguyen

AFTERWORD

I often joke with family and friends that I'm looking for redemption and salvation. As with everything that is inherently valuable, you can't get it from others, you can't buy it on sale in a value pack, and you definitely can't find it if you are not really looking for it. My family and I have been blessed by the kindness of people halfway across the world that we have never met. Their kindness forever changed the course of our lives. This is my attempt to pay it forward.

In our initial 8 weeks of fundraising at the end of 2020 – in the middle of a pandemic – we raised over $18,000 to provide educational aid and mentorship to kids. In an environment where a lot of us are struggling financially, emotionally, and maybe even physically, we are finding ways to come together and help those who need it the most. I am grateful to all of my family, friends, peers, and co-workers (present and past) who chose to support the mission of Winning the Hopes. THANK YOU!

Special thanks to Gennadiy Belenkiy, Karenann Terrell, Paul Milkman, and Ken Tanji for lending me their time and voices to shine a spotlight on Winning The Hopes through their personal interviews.

Here is a list of our inaugural donors. With all my heart, THANK YOU!

2020 Donors

Adam Young	James Burrows	Ned Demong
Amit Shukla	James Gazzola	Nhien Nguyen
Ash Laul	James O'Leary	Oz Akram
Becky Leung	Jared Walters	Paranjay Malhotra
Brijesh Varier	Jennifer Wang	Pasquale Iaconetti
Cabrina Attal	Jian Xue	Patrick Caruso
Charles Gazzola	John Akers	Patrick McCarthy
Chris Weiss	John Conroy	Patrick Uddenberg
Christina Tran	John Mohring	Philip Dugonjic
Colleen Jackson	John Nassar	R Rich
Dan Toppari	John Viele	Raymond Curbelo
David Clolinger	Jonathan Balkin	Richard Brounstein
David Izbicki	Jonathan Etkin	Richard Cmiel
David Roeser	Kenneth Tanji	Robert Kotz
Debbie Sherwin	Kevin Lemire	Scott Ingram
Drew Paradine	Kimba Nguyen	Sean Knipe
Eduardo Baez	Kurt Peterson	Sebastian Amaya
Eliot Meadow	Kyle Dufresne	Shany Biran
Emanuel Psyhojos	Lauren Rhode	Sharon Montreuil
Enrique Ybarra	Mark Donsky	Shea Baker
Erin Gondeck	Mary Coughlin	Simon Dyson
Geoff Hart	Melissa March	Sophia Sookram
George Mansour	Melvin Acevedo	Steven Tran
Greg Harnett	Michael Yee	Trang Nguyen
Hanyul Lee	Michelle Blunda	Tyler Lundberg
Henri Wajsblat	Mike Loftus	Will Liu
Jacqueline Chen	Natassa Boukouvala	Zaf Kamar

ACKNOWLEDGEMENTS

This book would not have been possible without the expert editing of Cabrina Attal and Kathy Clolinger. This is our third collaboration, and they are my favorite friends from when I called Chicago home. Now if I could just get them to be my friends on Facebook, I would have tripled my friend count. Last but not least, I wanted to thank my wife and rascal kids. They are the reason why I get up every morning.

Thank you, Pia Reyes, for the simple yet beautiful book cover design.

Thanks for reading my third book. Please add a short review on Amazon and let me know what you think!

ABOUT THE AUTHOR

Trong has more than twenty years of experience in sales, marketing, and consulting with the world's largest enterprises. Trong honed his craft at IBM, Dell and Microsoft. He earned his Bachelor of Arts degree in Economics from Western University – Canada, as well as an MBA from The University of Chicago.

Trong's first two books, *Winning The Cloud* and *Winning The Bank*, became underground hits with avid sales fans.

You can reach Trong at: https://winningthehopes.org

Trong Nguyen

WORKS CITED

Barber, N. (2009, December 1). *The Key To Happiness*. Retrieved from Psychology Today: https://www.psychologytoday.com/ca/blog/the-human-beast/200912/the-key-happiness

Bianca, V. L. (2018, June 23). *The Dangerous Romance of the " "Brilliant And Tortured Artist"*. Retrieved from Medium: https://medium.com/@Vi_LaBianca/the-dangerous-romance-of-the-brilliant-and-tortured-artist-8a31ad30411d

Consultants Mind. (2013, February 22). *3 kinds of power: positional, relational and expertise*. Retrieved from Consultants Mind: https://www.consultantsmind.com/2013/02/21/3-kinds-of-power-positional-relational-and-expertise/

Dalberg Advisors & Intel. (2016, June 23). *Decoding diversity: the financial and economic returns of diversity in tech*. Retrieved from https://www.dalberg.com: https://www.dalberg.com/our-ideas/decoding-diversity-financial-and-economic-returns-diversity-tech

Dettling, L., Hsu, J., Jacobs, L., Moore, K., & Thompson, J. (2017, September 27). *Federal Reserve Systems*. Retrieved from federalreserve.gov: https://www.federalreserve.gov/econres/notes/feds-notes/recent-trends-in-wealth-holding-by-race-and-ethnicity-evidence-from-the-survey-of-consumer-finances-20170927.htm

Dweck, C. (2006). *Mindset: The New Psychology of Success*. New York: Random House.

Erwin, M. (2019, August 1). *6 Reasons We Make Bad Decisions, and What to Do About Them*. Retrieved from Harvard Business Review Online: https://hbr.org/2019/08/6-reasons-we-make-bad-decisions-and-what-to-do-about-them

First Round Capital. (2019, January). *Eight Ways to Make Your D&I Efforts Less Talk and More Walk*. Retrieved from firstround.com: https://firstround.com/review/eight-ways-to-make-your-dandi-efforts-less-talk-and-more-walk/

Fottrell, Q. (2018, March 4). *Psychologists say they've found the exact amount of money you need to be happy*. Retrieved from Market Watch: https://www.marketwatch.com/story/this-is-exactly-how-much-money-you-need-to-be-truly-happy-earning-more-wont-help-2018-02-14

Hamermesh, D. (2011). *Beauty Pays: Why Attractive People are More Successful.* Princeton, New Jersey, USA: Princeton University Press.

Hardy, B. (2015, July 6). *The Secret To Happiness is Ten Specific Behaviors.* Retrieved from The Observer: https://observer.com/2015/07/the-secret-to-happiness-is-ten-specific-behaviors/

Heathfield, S. (2020, January 5). *What Makes a Bad Boss—Bad?* Retrieved from The Balance Careers: https://www.thebalancecareers.com/what-makes-a-bad-boss-bad-1917716

Hunt, V., Layton, D., & Prince, S. (2015, January). *Why Diversity Matters.* Retrieved from https://www.mckinsey.com: https://www.mckinsey.com/business-functions/organization/our-insights/why-diversity-matters

Johnstone, K. (2016, 03 29). *Should You Hire Attractive Salespeople?* Retrieved from https://www.peaksalesrecruiting.com: https://www.peaksalesrecruiting.com/blog/hire-attractive-salespeople/

Julson, E. (2018, May 10). *10 Best Ways to Increase Dopamine Levels Naturally.* Retrieved from Healthline: https://www.healthline.com/nutrition/how-to-increase-dopamine

Kan, J. (2018, August 13). *Why You Should (And Shouldn't) Join a Startup.* Retrieved from Atrium: https://www.atrium.co/blog/work-at-a-startup/

Klotz, A., & Bolino, M. (2019, July 31). *Do You Really Know Why Employees Leave Your Company?* Retrieved from Harvard Business Review: https://hbr.org/2019/07/do-you-really-know-why-employees-leave-your-company

Krommenhoek, B. (2018, April 10). *Why 90% of Startups Fail, and What to Do About It.* Retrieved from Medium: https://medium.com/swlh/why-90-of-startups-fail-and-what-to-do-about-it-b0af17b65059

Kununu. (2018, November 7). *What Makes Good Employees Quit? 16 Most Common Reasons.* Retrieved from Kununu: https://b2b.kununu.com/blog/why-do-good-employees-quit-leave-their-job

Maggiulli, N. (2020, June 2). *How Big Is The Racial Wealth Gap?* Retrieved from Of Dollars And Data: https://ofdollarsanddata.com/racial-wealth-gap/

Mayer, D., & Greenberg, H. H. (2006, August). *What Makes a Good Salesman.* Retrieved from Harvard Business Review: https://hbr.org/2006/07/what-makes-a-good-salesman

National Institute of Mental Health. (2019). *5 Things You Should Know About Stress.* Retrieved from https://www.nimh.nih.gov: https://www.nimh.nih.gov/health/publications/stress/index.shtml

National Philanthropic Trust. (2019). *Charitable Giving Statistics.* Retrieved from National Philanthropic Trust:

https://www.nptrust.org/philanthropic-resources/charitable-giving-statistics/

Parkin, S. (2018, March 4). *Has dopamine got us hooked on tech?* Retrieved from The Guardian: https://www.theguardian.com/technology/2018/mar/04/has-dopamine-got-us-hooked-on-tech-facebook-apps-addiction

Rain Group. (2019, August). *Discover what it takes to break through to executives, fill your pipeline, and win more sales.* Retrieved from Rain Sales Training: https://www.rainsalestraining.com/sales-research/sales-prospecting-research

Schultz, M. (2018). *Top Sales Prospecting Challenges.* Retrieved from Rain Sales Training: https://www.rainsalestraining.com/blog/top-sales-prospecting-challenges

Schwantes, M. (2018, December 21). *Why Do Employees Quit on Their Bosses? Because of 5 Common Reasons Still Not Addressed, Says New Research.* Retrieved from Inc.: https://www.inc.com/marcel-schwantes/why-do-people-quit-their-jobs-exactly-new-research-points-finger-at-5-common-reasons.html

Sheldon, K. (2012, 03 13). *8 Reasons To Choose A Startup Over A Corporate Job.* Retrieved from Fast Company: https://www.fastcompany.com/1824235/8-reasons-choose-startup-over-corporate-job

Sheridan, K. (2015, September 11). *19 Traits of Bad Bosses.* Retrieved from Association for Talent Development: https://www.td.org/insights/19-traits-of-bad-bosses

Stress Management Society. (2019). *How It Affects Us*. Retrieved from Stress Management Society: https://www.stress.org.uk/how-it-affects-us/

Tilt. (2019). *The Science*. Retrieved from Tilt 365: https://www.tilt365.com/The-Science

Wakeman, C. (2010). *Reality-Based Leadership: Ditch the Drama, Restore Sanity to the Workplace, and Turn Excuses into Results*. New York: Jossey-Bass.

Zara, C. (2012). *Tortured Artists: From Picasso and Monroe to Warhol and Winehouse, the Twisted Secrets of the World's Most Creative Minds*. New York: Adams Media.

Zenger, J., & Folkman, J. (2014, September 1). *9 Habits That Lead to Terrible Decisions*. Retrieved from Harvard Business Review Online: https://hbr.org/2014/09/9-habits-that-lead-to-terrible-decisions

Zetlin, M. (2018, October 5). *Why Is a Startup Worth $1 Billion Called a Unicorn? The VC Who Invented the Term Explains*. Retrieved from Inc.: https://www.inc.com/minda-zetlin/unicorn-1-billion-valuation-vc-venture-capital-aileen-lee-cowboy-ventures-elite-universities.html

www.ingramcontent.com/pod-product-compliance
Lightning Source LLC
Chambersburg PA
CBHW060027210326
41520CB00009B/1024